BASIC BIOINFORMATICS

BASIC BIOINFORMATICS

S. Gladis Helen Hepsyba

Lecturer
Department of Zoology
Dr. Ambedkar Goverment Arts and Science College
Chennai

C.R. Hemalatha

Lecturer
Department of Bioinformatics
Sri Ramachandra University
Chennai

MJP PUBLISHERS

ISBN 978-81-8094-386-7

MJP Publishers
No. 44, Nallathambi Street,
Triplicane, Chennai 600 005

MJP 052

Publisher : C. Janarthanan

PREFACE

Bioinformatics entails the creation and advancement of databases, algorithms, computational and statistical techniques, and theory to solve problems arising from the analysis of biological data. Rapid developments over the past few decades in molecular research technologies and developments in information technologies have combined to produce tremendous amount of information related to molecular biology. It is the name given to these mathematical and computing approaches used to glean understanding of biological processes. Here, computers are used to gather, store, analyse and integrate biological and genetic information which can then be applied to gene-based drug discovery and development.

It seems imminent that biology and computers are helping and influencing each other and synergistically merging, to become one. The need for bioinformatics capabilities has been precipitated by the explosion of publicly available genomic information resulting from the Human Genome Project and the like.

This introductory text intended for undergraduate and postgraduate courses links issues in computer science to biology to offer a clear picture of the principles driving bioinformatics. It strikes an emphasis on the ideas underlying the basis of biology essential for the students of bioinformatics. The introductory chapter provides information to students to grasp computer and Internet usage forming a platform to access the available resources using the Internet.

DNA sequencing techniques, the first step to obtain data for any bioinformatics analysis is described precisely. It starts with the conventional methodologies and concludes with the latest ones in the field. This book also opens the gateway to enhance understanding of the arrangement of the nucleotides sequenced inside a genome by elucidating various procedures on mapping of genomes.

Database concepts, biological database management systems and designing a biological database, being the brain of all bioinformatics work, have been clearly elucidated throwing light on the various types of databases a bioinformatician would be interested in and also brief descriptions of the most important ones.

Consequent chapters discuss various types of sequence analysis, software used for the same, advantages and disadvantages of using each software, etc. It will enable students to understand how much the As, Ts, Gs and Cs in a DNA sequence make sense and also drive the process called life, equipping students in identifying and understanding the evolutionary relationships among the many different kinds of life on earth, both living (extant) and dead (extinct). Applications of bioinformatics, which are indirectly opening new avenues of research, have been explained from a genomics and proteomics perspective.

We sincerely wish that all our readers will have an enjoyable experience, traversing through the basic principles driving an emerging science, bioinformatics.

S. Gladis Helen Hepsyba

C. R. Hemalatha

ACKNOWLEDGEMENTS

Of all powers, supreme is the power of the Almighty. We are indeed grateful to God, for having been able to impart the knowledge that we have acquired in the form of a book.

Gladis wishes to thank Prof. V. Prabhakaran, Head, Department of Zoology, Dr. Ambedkar Government Arts and Science College, for his encouragement throughout the period of this work. She also appreciates the support rendered by Mr. K. Rajkumar, Librarian of Dr. Ambedkar Government Arts and Science College, Chennai. She is also grateful to her husband Mr. C. David Sam Titus for his moral support in all her endeavours.

Hemalatha wishes to thank Dr.K. Anand Solomon, Asst. Professor, Department of Bioinformatics, Sri Ramachandra University for his unflinching encouragement, constant support and invaluable suggestions at each stage. She is also grateful for the help rendered by Mrs. Deepa Parvathi, Lecturer, Department of Human Genetics, Sri Ramachandra University and Dr. P.L. Sujatha, Asst. Librarian, Department of Bioinformatics, Madras Veterinary College, for taking time off their schedule to read the manuscript and offer suggestions for improvement. She appreciates and thanks the painstaking work of typing done by Ms. Anandhi Iyappan and Ms. S. Dhaarini, postgraduate students, Department of Bioinformatics, Sri Ramachandra University.

Throughout our endeavour to bring out this book, our families have stood by us with overwhelming love and unfailing support, sacrificing a lot in the process. We cannot thank them enough for all their encouragement, understanding and prayers now and always.

We owe our profound thanks to Mr. J.C. Pillai of MJP Publishers for having accepted to publish this work. We are also thankful to Mr. C. Sajeesh Kumar, Mrs. P. Parvath Radha and the entire team for having shaped the manuscript into what it is today.

S. Gladis Helen Hepsyba
C. R. Hemalatha

CONTENTS

1

INTRODUCTION

The mathematical, statistical and computing methods that aim to solve biological problems using DNA and amino acid sequences and related information is classical bioinformatics.

Fredj Tekaia

OBJECTIVES

1. To provide background information which will help demystify computer usage
2. To provide an introduction to the resources available to biologists using the Internet in sufficient detail to allow them to explore and learn how to use these resources on their own
3. To provide practical instructions to the students on using the specific network resources required

When molecular biologists started to generate DNA sequence data, years ago, it was natural that computer scientists and mathematicians would take a keen interest. The analog world of biology was digital information: a linear string of four chemical groups encoding the entire blueprints for the protein machinery of the living cell. This field of study gained a real identity and the name "Bioinformatics", in the mid-1980s. This was the dawn of bioinformatics, the intersection of molecular biology and computer science. Integrating information in the molecular biosciences involves more than the cross-referencing of sequences or structures. Experimental protocols, results of computational analyses, annotations and links to relevant literature form integral parts of this information, and impart meaning to sequence or structure.[1]

Even though the three terms—bioinformatics, computational biology and bioinformation infrastructure—are oftentimes used interchangeably, broadly, they may be defined as follows:

- *Bioinformatics* refers to database-like activities, involving persistent sets of data that are maintained in a consistent state over essentially indefinite periods of time
- *Computational Biology* encompasses the use of algorithmic tools to facilitate biological analyses
- *Bioinformation Infrastructure* comprises the entire collection of information management systems, analysis tools and communication networks supporting biology.

Thus, the latter may be viewed as a computational scaffold of the former two.

The translation of DNA sequences into biologically meaningful information has given birth to a new field of science called bioinformatics. The term has been derived by the contribution of biology and informatics, i.e., applied information science (mathematics, statistics and computer science) to increase the understanding of biology. It is a multidisciplinary science consisting of three core areas, namely molecular biology databases, sequence analysis and emerging technologies like microarrays.

Digitalization of data has reached a high level however, information is still dispersed. Individual and coordinated approaches have been initiated to improve accessibility of biological resource centres, their holdings and related information through the Internet. These approaches cover the subjects such as standardization of data handling and data accessibility, and standardization and quality control of laboratory procedures.[2]

The first step is data representation. The DNA is not neatly arranged in the pristine double helix that we all recognize. It is coated with proteins that bind to specific sequences, which untwist the helix to allow gene expression and wind it up into tightly packed supercoils. Far from being a static archive of blueprints, DNA is a complex, dynamic, three-dimensional molecule and yet we represent all of this as a simple string of the characters A, C, G and T.

Second is the concept of similarity. Evolution has operated on every sequence that we see today. It conserves genes that encode important proteins and sequences that are involved in gene regulation. Sequences that encode useful functions are transferred, like code modules, from one organism to another. Because of evolution, similar sequences have similar functions. Algorithms for comparing sequences and finding similar regions are at the heart of bioinformatics. At many different levels, they are used to find genes, determine their functions, study their regulation and assess how they, and entire genomes, have evolved over time.

Third is the reality that bioinformatics is not a theoretical science; it is driven by data, which in turn is driven by the needs of biology. Relatively few researchers have the luxury to develop algorithms and theories in the traditional academic sense, while most of them are consumed in the management and analysis of data. The introduction of automated DNA sequencing in the early 1990s created what was, at the time, a torrent of sequence data; but it was the Human Genome Project (HGP), with its massive automation which really opened the floodgates in the past few years.

SOFTWARE IN BIOINFORMATICS

Two main factors have shaped the current landscape of bioinformatics software. As already mentioned, the field has been driven by the massive amount of data and the research projects that generate it. As a result, most people in bioinformatics work on very focused projects and few have the luxury to sit back and write the ideal program for gene prediction. In addition, the technologies used in the lab and the data they produce, have evolved very rapidly. That has made it very difficult to commit a lot of resources and time to specific pieces of software. The lifespan of a software project is often quite short and the lead time before deployment is minimal. Being able to understand the essence of a problem and hack up a quick solution that gets the job done are critical skills for a good bioinformatics developer.

A classic example is the *Genome Assembler* written by Jim Kent at UC Santa Cruz. Excellent software already exists for assembling the fragments of data produced by sequencing instruments into large blocks but it could not handle the scale of the task that the human genome project had created. Rather than trying to modify the existing code, it made sense for Kent to start from scratch and build something, in very short order, that was tailored to the task at hand. More than a quick hack, but a lot less than a complete, polished product, Jim's software assembled the human genome.

SCOPE AND APPLICATIONS

Research in bioinformatics includes methods of development for storage, retrieval, and analysis of the data. Bioinformatics is a rapidly developing branch of biology and is highly interdisciplinary, using techniques and concepts from informatics, statistics, mathematics, chemistry, biochemistry, physics, and linguistics. It has many practical applications in different areas of biology and medicine. It has emerged as a distinct discipline that straddles the interface between the traditional biological sciences, the computer sciences and advanced computational methodology. It is rapidly becoming a powerful new approach to understand life and it may well reverse the reductionist paradigm that has held sway in molecular biology ever since Erwin Schrodinger turned on a generation of physicists to biology with his publication of *What is life*? more than 50 years ago.

The new discipline of bioinformatics promises to provide the tools needed to attack the complexity of conducting holistic biological research. According to Delici, "... because of these complexities, biology will eventually become the most computational science, surpassing physics," and predicts that within the next 10 to 15 years, bioinformatics will become an integral part of biology.

The biotechnology and drug discovery industries have witnessed the development of high-throughput automated equipment, which enable amassing of data faster than it can be analysed and utilized. Pharmaceutical research will clearly be the major benefactor of development in bioinformatics. Hundreds of new drug targets have been identified by using computational techniques, which involves searching for genes similar to known protein-encoding genes. In the future, virtual toxicology screening may be the first set-up in predicting the effects of new chemical or complex metabolic pathways. In addition, bioinformatics will likely provide the methodology, to make accurate predictions about protein tertiary structure based on amino acid sequences and a viable means to design drugs based on computer simulation of the docking of small molecules to the predicted protein architecture.

Whole genome analyses and sequences, experimental analyses involving thousands of genes simultaneously, DNA chips and array analyses, comparative analyses of species and strains, proteomics analysis, forensic medicine and agriculture are some of the areas where bioinformatics may be applied. A number of recent workforce studies have shown that there is a high current and unmet demand for people trained to various levels of expertise in bioinformatics, to serve the upcoming biotechnology and biopharmaceutical industries which have observed significant growth in the genomic era.

The completion of the human genome project and other genome sequencing projects has spearheaded the emergence of the field of bioinformatics. Using computer programs to analyse DNA and protein information has become an important area of life science research and development. While it is not necessary for most life science researchers to develop specialist bioinformatics skills (including software development), basic skills in the application of common bioinformatics software and the effective interpretation of results are increasingly required by all life science researchers. Training in bioinformatics is increasingly occurring within the university system as part of existing undergraduate science and specialist degrees.[3]

ELEMENTARY PROTOCOLS

The Internet has become an important tool for biological and biomedical research scientists. Using the Internet, it is possible to perform a number of analyses on research data and to search for and obtain information. Over the last several years, the number of tools and the amount of information relevant to biologists and available on the Internet has grown and the ease of use of these tools has grown as well. As a result, the value of Internet resources for biologists now significantly outweighs the costs in time and money of using it.

Internet Services

The various Internet services available are characterized by client software, server software, a set of capabilities they agree upon (e.g. text, pictures in the GIF or Graphic Interface Format, etc.), a protocol by which they communicate (i.e., how the data is encoded in a stream of bytes) and a port on which the client contacts the server to be communicated. The distinction between different services can be blurred because, different services can perform similar functions, different services can share capabilities and also because of the existence of multifunction clients and servers. Specifically, many modern web clients are highly multifunctional, namely Gopher, File Transfer Protocol (FTP), E-mail and net news clients and web clients.

Telnet Telnet is one of the oldest network services and perhaps the easiest to understand. It allows one computer to log on to another computer as if it were a terminal. Once logged on, you will have all the privileges of a local user, you can run programs, create and delete files. This is probably the most common way that users with accounts will use a computer. A telnet session can negotiate a range of different protocols but this almost always includes ASCII (American Standard Code for Information Interchange) text, because many protocols are used for other services (e.g. to connect to a server for other protocols). It is possible to connect to a web server with a telnet client, if we understand the syntax of HTTP. This is almost never done to use a web server, but is occasionally done when debugging (removing computer mistakes from a program). Telnet is useful for interactive computer access, but is much less useful for transferring files. From a practical point of view, every telnet host will be different, and thus we will need to learn about each one as and when an occasion arises.

File transfer protocol (FTP) FTP is an older Internet service designed specifically for file transfer. Originally like telnet, it was intended for account owners. However, as it became apparent that it was useful to make files available to the world at large without giving all those who wanted an account, the variant of anonymous FTP developed. In this variant, logging in with a magic user name (most commonly anonymous or FTP) eliminates the requirement for a password. In 1996, the use of the World Wide Web (www) has rendered FTP access obsolete. Although there was and is some truth to that statement (especially given that files on an FTP server can be retrieved by a web client), the need for FTP clients has not vanished. Some users will choose to avoid them, preferring the simplicity of dealing with a single piece of software, but within their domain. FTP clients are more versatile than web browsers. In some cases one has more control with an FTP client, and for simple file transfers they are quicker and more convenient.

E-mail Both FTP and telnet are interactive, more or less real-time programs. Sometimes it is useful, however, to communicate with another computer, or more commonly, a user on another computer, by leaving them a message, which they can read and respond to at their convenience. This is done over the Internet by using e-mail. E-mail is a generic term for a variety of processes which can use different protocols and network technology and which, in many cases uses a more complex client/server model than many of the other protocols discussed in this chapter. At present, most e-mails are transmitted by SMTP (Simple Mail Transfer Protocol) via TCP/IP (Transfer Control Protocol/Internet Protocol) over the Internet.

Using the web Words, pictures, hyperlinks between and within documents, movies, music, online shopping, database access, and even basic attempts at virtual reality—there is nothing the web cannot do. The web describes information using Hypertext Markup Language (HTML) and transmits it using Hypertext Transfer Protocol (HTTP). The current common name, the web, is a contraction of its original name, the World Wide Web, also abbreviated as www or w3. A web browser performs multiple tasks. First, any web browser is an HTTP client, it knows how to transfer data using the HTTP protocol.

Second, any web browser also knows how to interpret and display HTML, the content markup language used on the web. Different browsers have different display capabilities and display the same HTML code in different ways (which is why HTML is referred to as a content markup language instead of a page description language) but all of them can understand (parse) HTML and do something reasonable with it.

Some of the differences in the way different web browsers display the same web page come from different design decisions ("what font should be used for <H1> text?"), and some of it comes from the fact that different web clients have different capabilities. Some of these differences, such as the ability to display various kinds of still or moving images as part of the Web page or to run programs written in Java, Active X, or JavaScript, represent extensions to HTML. These extra capabilities may be built into the browser or may be added by plug-ins (software extensions which give the browser new functionality). The behaviour of a web browser can be frequently changed by configuring its preferences, if you find the default font too small, it can be increased.

Many new computer users assume that the web and the Internet are synonymous. However, many protocols other than HTTP flow over the Internet. In part, the new user is confused by the fact that, in addition to supporting extensions to HTML, many popular web browsers have support for other protocols such as e-mail [SMTP, POP (Post Office Protocol), IMAP (Internet Message Access Protocol)], newsgroups, FTP, and Gopher. What this really means is that the particular piece of software (e.g. Netscape Communicator) is more than just a web client; it is also an e-mail client, an FTP client and a Gopher client. Finally, HTTP does not have to be transmitted over the Internet and HTML does not have to be transmitted via HTTP. Web technology has become a common interface tool for communication between computers on a local network (Intranet). Because web clients are also a limited FTP client, many people choose to use them. In case a web page contains a link to an FTP server, by simply selecting the link the file may be downloaded.

SEARCH ENGINES

The web is vast and disorganized and the overwhelming majority of what is there is irrelevant to us. Further, the web changes constantly; new resources appear, old resources become outdated or disappear and the paths and techniques used to access resources change. The bad news is that there is no perfect way of finding the resources that might be useful to us. The good news is that we do not need a perfect way; finding even half the useful resources on the web is well worthwhile and a lot better than ignoring the web altogether.

There are several websites from which you can launch searches. You type in words that you expect to find on the web page you are looking for and it will return to you a list of pages on the web that contain those words. Google is one of the finest search services which provides a much shorter, more focused list of sites than most other services. Alta Vista is a very complete and powerful search facility.

When compared to Google, Alta Vista frequently generates a usually large list of sites. Other search engines give different results. If one of the above two search services does not give satisfactory results, Infoseek, HotBot, Yahoo, MSN, etc. may be used.

Natural Links

The original dream of hypertext was that natural relationships between topics would be linked. Within the realm of biology, this dream is increasingly coming true. Although such links are not nearly as complete or as easy to use, they are improving and already are extremely useful. If one accesses Medline via Entrez, for example, one may link from the bibliographic citation retrieved from Medline to a full text copy of the article on the publisher's website. Similarly, one can link from OMIM (Online Mendelian Inheritance in Man) to the HUGO (Human Genome Organization) gene nomenclature site to GeneCards to Tumour Gene Database. Such links are not always labelled in the most obvious ways or located on a page or within a site exactly where one would expect them to be for the present. It is important to explore a site creatively in order to find all the valuable links that might be there.

PROBLEMS ON THE WEB

One of the major problems currently afflicting the web is that of incompatibility. This is particularly unfortunate and ironic in that one of the goals of the original web development is seamless integration of information resources across the Internet. One common symptom of this problem is a browser logo and a statement like "Best Viewed with Netscape 4.5". One attempt at solving this problem is the campaign for platform-independent websites.

Another recurring problem is that of security. It is good that one keeps in touch with the author of any browser and upgrade to new versions as warranted. It is better that we consider turning off support for Java in our browsers and above all, be cautious, alert and informed of all pros and cons of web usage.

Some Practical Considerations for Using the Web

1. It has become increasingly possible to spend money on the web. Nonetheless, with one exception, it is unlikely that we will spend money by accident. For example, while using the otherwise free service Entrez we might innocently link to and download a SWISS-PROT sequences file. According to Geneva bioinformatics, we have just become liable for an annual licencing fee and there is no way we could have known that. Thus, if you are working for a for-profit organization, we should either carefully avoid SWISS-PROT or else pay the licencing fee. Other than that, we will find that many resources on the web are free and those that are not, inform us clearly and carefully before obligating us to pay.

2. Although the Internet is becoming increasingly important to biologists, it is still not a sufficient resource for keeping up with biology. With access to a good (or even adequate) science library, we can do without the Internet. This is changing. Reasonable people might differ as to whether the library or the web is more important but most would probably agree that both are now virtually essential for the working biologist. To be clear, physical access to a library is becoming less essential but belonging to an organization affiliated with a library remains important. Most research is still published in conventional journals. What has changed is that a complete copy (full text and figures) of more and more journals have become available via the web. However, most of these are only available to subscribers.
3. Network resources are not as reliable as one would like. If we select a resource and receive only an error message, the fault is likely to be with the server. It is even the case that if we reach the server and do not receive the results we expect (e.g. search for a common term and get nothing in return) this might be due to the fact that the server is showing error rather than we doing the search incorrectly. And finally, what has become an increasing problem is that web pages will either move, be edited to significantly change their meaning or become temporarily or permanently unavailable.
4. Compared to the well-developed conventions for referencing data from the paper literature, conventions for acknowledging of Internet resources in a thesis or

paper are in utter chaos. Surfing the net for a few months will uncover a number of competing standards for how to accomplish this. For resources available on the web, a Uniform Resource Locator (URL) is a good reference.

5. Just because it is on a computer it does not mean it is correct. This sounds like a banal truism, but it is startling how good the beauty of computer output can make bad data look. Genbank is loaded with author errors (both in the sequences and comments sections) and probably contains some archivist-introduced errors as well. The same is true of Medline. Any study, which assumes perfection in such data, will produce an incorrect result.

This introductory chapter offers a clear exposition of the principles driving bioinformatics. Being accessible to both biology and computer science students, it demonstrates that relatively few techniques can be used to intuitively solve a large number of practical problems in bioinformatics.

REVIEW QUESTIONS

1. What is the scope of bioinformatics? Enlist the applications of the same.
2. Describe in detail about the various types of Internet services.
3. Discuss the role of search engines in bioinformatics. Add a note on Medline.

REFERENCES

1. Alexander Garcia Castro, Yi-Ping Phoebe Chen and Mark, A. Ragan. (2005). "Information integration in molecular bioscience." *Appl. Bioinformatics.* 4(3): 157–173.
2. Romano, Paolo; Kracht, Manfred; Manniello, Maria Assunta; Stegehuis, Gerrit; Fritze, Dagmar. (2005). "The role of informatics in the coordinated management of biological resources collections." *Applied Bioinformatics.* 4(3):175–186.
3. Sonia Cattley and Jonathan, W. Arthur. (2007). "BioManager: the use of a bioinformatics web application as a teaching tool in undergraduate bioinformatics training." *Briefings in Bioinformatics.* 8(6):457–465.

2

DNA SEQUENCING TECHNIQUES

Sequencing the chimp genome is a historic achievement that is destined to lead to many more exciting discoveries with implications for human health.

Francis Collins

OBJECTIVES

1. To understand the biochemical methods for determining the order of the nucleotide bases
2. To draw differences between the conventional and modern methods of DNA sequencing

INTRODUCTION

The central dogma of molecular biology tells how genes represented by DNA sequences are copied to mRNA, which is then translated into functional proteins. DNA sequence contains the instructions for everything a cell does, from the moment of conception until death. Knowing the complete DNA sequence facilitates understanding of the molecular components of the cell and their interactions.

Determination of the order of nucleotides in the DNA molecule (RNA or proteins also) is called as sequencing.

The two important techniques for DNA sequencing are—the enzymatic chain termination method (Sanger–Coulson method) and the chemical degradation method (Maxam–Gilbert method). Both will generate nested sets of single-stranded DNA fragments, which are separated by size on an acrylamide gel. When compared to Maxam–Gilbert method, the Sanger–Coulson chain termination method generates more easily interpreted raw data and has become the most widely used method. An important step in large-scale sequencing was the development of automated DNA sequencers.

The preferable approach for sequencing of large genomes is still under discussion. Obviously, directed strategies can only be considered if a physical map has been or will be constructed. The approach for many sequencing efforts on large eukaryotic genomes including *S. cerevisiae* and *C. elegans* has been under construction of overlapping arrays of large insert clones, followed by complete sequencing of these clones one at a time, either by shotgun or directed strategies. Similar strategies have also been applied for smaller bacterial genomes like *M. pneumoniae*.

In these cases, genome sequencing has been initiated after the construction of a physical map. Shotgun sequencing of whole genomes was first applied at The Institute of Genome Research (TIGR), then headed by Craig Venter. In 1995, the 1.83-Mb genome of *Haemophilus influenzae* was completely determined by an assembly of 24,000 random sequence reads from plasmids with inserts followed by directed approaches on lambda clones and PCR products to fill the remaining gaps. The overall redundancy was 6.3 and the quality was estimated to 1 error in 5000 ± 10000 bases. Very similar emphasis laid on lambda clones with large inserts (16 kb) in parallel to the plasmid sequencing. By including these sequences in the assembly process, the lambda clones were automatically ordered and thus, a physical map was obtained. Several other genomes of bacteria and archaea have been sequenced using the whole genome shotgun approach. Sequencing of the human genome has for many years awaited the sequences of the model organisms but is now completed fully. This involves both clone-by-clone approach as well as shotgun approach.

cDNA SEQUENCING

The ultimate goal for sequencing of whole genome is to decode all the genetic information carried in the genome. However, there is a drastic variation in the amount of useful information among different organisms. Bacterial genome shotgun approaches also carry compact genetic information with up to 88% coding sequence (*M. genitalium*). In addition, the sequence is easily interpreted due to lack of introns. Sequencing of larger eukaryotic genomes is less rewarding. The coding sequence in the human genome represents only approximately 3% of the total DNA and for some plant genomes, the figure is even lower. Eukaryotic gene identification by whole genome sequencing is therefore slow and coding regions are difficult to predict due to the introns.

A more efficient method for gene identification in eukaryotic genomes is sequencing of Expressed Sequence Tags (ESTs). ESTs are partial sequences of cDNA, reversibly transcribed from mRNA and represent a direct supply of coding intron-free sequences of genes. Even if techniques for enrichment of full-length clones have been adapted, many transcripts lack sequence in the 5′ end due to internal priming or due to internal restriction sites. To specifically determine missing 5′or 3′ sequences, a variety of methods based on the polymerase chain reaction have been developed. Rapid amplification of cDNA ends (RACE) was first described, closely followed by the very similar anchored PCR (A-PCR) technique. To find missing 3′ ends the first strand synthesis is performed with an oligo(dT) tail and a gene-specific primer, designed from the partial cDNA sequence selected (i.e., ESTs or fragments selected by differential approaches). For isolation of the 5′ end, PCR can be performed with the specific primer and a primer annealing to the new primer site. Originally the new primer site was introduced by homopolymer tailing using terminal transferase, but greater success has been obtained by using anchor oligonucleotides and T4 RNA ligase. Biotin capture of the gene-specific first strand fragments has also been used.

The throughput in full-length sequencing does not match the speed for which new gene candidates are produced by EST efforts or differential expression techniques. Yet full-length and high-quality sequences can potentially provide more accurate database comparisons as well as enhance the ability for different computer programs to predict gene function and structure. Each selected clone is usually sequenced by primer walking, which is a well established but slow technique. Concatenation of cDNA sequencing (CCS) has been suggested as an alternative method of sequencing.

ASSEMBLY

Sequence assembly is a process that involves comparison of sequences, finding overlapping fragment pairs, merging as many fragments as possible and creating a consensus sequence from the merged fragments. Accurate assembly algorithms are essential for reconstruction of the original DNA sequence (cosmid, BAC, prokaryotic genome, cDNA clone) in shotgun sequencing and for grouping of ESTs in expression profiling.

A commonly used program for cosmid scale assembly is GAP4, which is one of several programs in the Staden package. It has a highly interactive graphical interface with several tools for manipulation and display of data.

The Phrap assembly program has been successfully used to assemble larger DNA fragments. It works especially well along with Phred, which is an improved lane tracking and base calling software and Consead, which is the graphical interface. The TIGR assembler was developed to manage a whole genome shotgun assembly, i.e., to assemble genomes of 0.5 + 4 Mb.

Such large assemblies means a dramatic increase in the number of pairwise comparisons required, an increased likelihood to encounter repetitive regions and a higher probability to obtain false overlaps due to chimeric clones. The TIGR assembler algorithm has been used for assembly of several prokaryotic genomes as well as for grouping of ESTs for gene indexing purposes. Assembly algorithms are applied on EST sequences to generate clusters of sequences deriving from the same transcript. It is a way to reduce the large quantity of EST data to a number of groups with overlapping sequences representing unique genes.

Among the efforts to form such gene indexes can be mentioned, Unigene, TIGR's, Human Gene Index (HGI) and Gene Express. The strategies for grouping of the sequences differ in stringency. The TIGR Assembler uses HGI, has a stringent gene-matching criterion, which prevents chimerism. On the other hand, the strictness results in a more fragmented representation, which disallows divergent ESTs, that represents alternative form of the same gene to fold into the same index class. Splice variants are accepted only if they match fully sequenced genes with known isoforms in the Expressed Gene

Anatomy Database (EGAD), which is a database with well-characterized human genes. Unigene and gene express represent very loose gene indices. Sequences are grouped into common classes if they share or overlap over a certain threshold, using BLAST, FASTA and Smith–Waterman methods for comparison. A single index class can then contain several splice forms of the same gene but chimeras and other artefacts may be incorrectly included.

There are several aspects that challenge a successful assembly. First, the error frequency in sequence raw data, which depends on experimental aspects like template, sequencing enzyme, sequencing chemistry and instrumentation, and aspects like tracking and base calling software. A general estimate for the quality of raw data can be obtained from EST sequences, which represents poorly edited, single-pass sequence reads. The overall accuracy for EST is usually about 97%. The error types include additions, deletions and substitutions of bases and are usually more abundant in the end of sequence. Other problems that occur are chimeric clones and lane tracking errors and for EST assembly, splice variants, clone reversals and internal priming errors.

In the assembly of ESTs, genes belonging to the same gene family can be hard to distinguish. This is especially difficult in plants, where as many as 20% (*Arabidopsis*) of the genes belong to gene families. The higher sequence diversity in the 3′-UTR (UnTranslated Region) of the transcripts is best used to distinguish between gene family members. Assembly of genomic clones however, meets problems in repetitive regions. This is especially difficult for organisms with very high GC content like *Deinococcus radiodurans* (68% GC) or high AT content like *Plasmodium falciparum* (82% AT). A correct assembly over long repetitive regions has to be performed with careful respect to the physical map or PCR fragments spanning the regions.

Genome Sequence Assembly using Trace Signals and Additional Sequence Information

Today large-scale genome sequencing efforts produce enormous quantities of data. They are now based on the chain termination dideoxy method.[1] Gel or capillary electrophoresis used can determine only about a maximum of 1000 to 1500 bases. The DNA base pairs have low error probabilities for the called bases often being around the first 400 to 500 bases. Current sequencing strategies for a contiguous DNA sequence (contig) ranging anywhere between 20 kilo bases (kb) and 200 kb. These will therefore basically boil down to fragment the given contig in hundreds or thousands of overlapping subclones, analyse these by electrophoresis and subsequently assemble the subclones back together in one contig.[2]

In fact that DNA tends to contain highly repetitive stretches with only very few bases differing across different repeat locations, impedes the assembly process in an awesome way.

The above-mentioned error rates and repetitive properties of DNA lead to the necessity of using fault-tolerant and alternative-seeking algorithms. The assembly problem even using error-free representations (fragments) of the true sequence is incomplete. This means that the volume of data can only be assembled by approximating strategies, relying on algorithms that are well behaved in time and space complexity.

Assembly Strategies

A sequence assembly is essentially a set of contigs, each contig being a multiple alignment of reads.[2] A number of different strategies have been proposed to tackle the problem, ranging from simple greedy pairwise alignments sometimes using additional,[3] sometimes using a whole set of refinements[4] to weak methods like genetic algorithms.[5, 6 & 7]

Nowadays there are mainly two different existing approaches for assembling sequences:

i. The iterative and
ii. The "all in one step" approach.

The first type of assembly is essentially derived from the fact that the data analysis and reconstruction approximation algorithms can be parameterized differently, ranging from very strict assembly of only the highest quality parts to very "bold" assembly of even the lowest quality stretches. An assembly starts with the strict parameters, having the output edited manually (by highly trained personnel) or by software and then the process is reiterated with less strict parameters until the assembly is finished or the parameters become too lax. The second approach has been made popular by the Phrap assembler presented by Phil Green (http://www.phrap.org/). This assembler uses low- and high-quality sequence data from the start and generates a consensus by puzzling its way through an assembly using the highest quality parts as reference, giving the result to a human editor for finishing.

A common characteristic to all existing assemblers is that they rely on the quality values with which the bases have been attributed by a base caller. Within this process, an error probability is computed by the base caller to express the confidence with which the called base is thought to be the true base. The positive aspect is the possibility for assemblers to decide in favour of the best, most probable bases when a discrepancy occurs. The negative aspect of current base callers is their inability to write confidence values for optional, uncalled bases at the same place. This could improve the search for alternative assemblies substantially.

Researchers therefore implemented a third type of assembler, trying to combine and substantially extend the strengths of both approaches mentioned above, and copying assembly analysis strategies done by human experts. An important criterion in the design of their assembler is the quality aspect of the final result; the assembler works only with the stretches of DNA sequences marked as 'high' or 'acceptable' quality, from which it selects the best stretch to start an assembly. These High Confidence Regions (HCR) ensure a firm base and good building blocks during the assembly process. Lower quality parts or Low Confidence Regions (LCR) can be used later on, if needed.

The main difference of their approach is that, they combined the assembler with some capabilities of an automatic editor. Both the assembler and the automatic editor are separate programs and run separately, but they view the task of assembly and finishing being closely related enough for both parts to include routines from each other. In this process, the assembler gains the ability to perform signal analysis on partly assembled data, which helps to reduce misassembles especially in problematic regions like repeats (*Alu* sequences, repeats, etc.), where simple base qualities alone cannot help. Analysing trace data at precise points with a given hypothesis in mind, there is a substantial advantage of a signal analysis aided assembler compared with a "sequential base caller and assembler" strategy, especially while discriminating alternative solutions during the assembly process. In return, the automatic finisher gains the ability to use alignment routines provided by the assembler.

Methods and Algorithms

They worked out a multiphase concept to have their assembler perform the difficult task of shotgun sequence alignments. Different authors have proposed different sets of acceptance criteria for an optimal alignment. Traditionally, the objective of this (assembly) problem has been to produce the shortest string that contains all the fragments as substrings, but in case of repetitive target sequences, this objective produces answers that are over-compressed. As a result of this, they conceived the strategy of the "least number of unexplainable errors" present in an assembly to be optimal.

To demonstrate their working principles, they used a toy project of six reads, which they assume to have an overall fair base quality.

Data Preprocessing

A HCR of bases within every read is to be selected as an anchor point for the next phases. Existing base callers (ABI, Phred and others) detect bases and rate their quality quite accurately and keep increasing in their performance, but bases in a called sequence always remain afflicted by increasing uncertainty towards the end of a read. This potential additional information can nevertheless constitute an impeding moment in the early phases of an assembly process, bringing in too much noise or preventing the correct determination of true, long-range repeats.

Another important factor is the sequencing vector, which will invariably be found at the start of each read. This earliest part of any cloned sequence must be marked or removed from the assembly. In analogy to the terms used in the GAP4 package, they will refer to LCR also as "hidden" data whereas, sequence marked as being part of the sequencing vector is "marked" data.

The following is the information the assembler will work with, any of which can be left out (except sequence and vector clippings) but will reduce the efficiency of the assembler.

1. The initial trace data, representing the gel electrophoresis signal

2. The called DNA sequence
3. Position-specific confidence values for the called bases of the DNA sequence
4. A stretch of DNA in each sequence marked as HCR
5. General properties like name of the sequencing template, etc.
6. Special DNA properties in different regions of a read (like sequencing vector, standard repeat sequence, etc.) that have been tagged or marked.

The data preprocessing step has been taken out of the actual assembler as almost every laboratory has its own means to define 'good' quality within reads and already use existing programs to perform this task. For example, quality clipping, sequencing vector and cosmid vector removal are controlled by the PREGAP script provided with the GAP4 package or can be done with Crossmatch4 provided by PHRAP.

Read Scanning

A common start for an assembly is to compare every read with every other read (and its reversed complement) using a fast and fault-tolerant algorithm to detect potential overlaps. Researchers developed an algorithm based on the Shift AND text search algorithm introduced by Wu and Manber which extended ideas of Baeza Yates and Gonnet in 1992. The Shift AND algorithm allows with a configurable error threshold, two sequences to be recognized as partly identical only by shifting bit vectors.

The type of errors within the partial identity is irrelevant as insertions, deletions and mismatches in sequences are equally recognized. Originally, the complexity of the Shift AND algorithm allowing errors is 0 *cmn* where, *c* is a constant depending on the number of errors allowed in the match, *m* being the length of the pattern and *n* is the length of the sequence to be compared with.

The efficiency of the algorithm has been demonstrated by testing it on simulated shotgun data sets of real world DNA sequences and comparing results. They found that the fast scanning method is especially well-suited for finding weak overlaps or overlaps in error-rich regions. For example, the DNASAND algorithm was always able to find potential overlaps with a false positive rate below 0.5 and with less than 2 per cent of missed overlaps, even in data sets with an artificially introduced high error rate of 10 per cent. Although the DNASAND algorithm does not specify the overall type of global relationship of two sequences (total correspondence, containment and overlapping), any type of this relationship is recognized. As a result of this first scan, a matrix containing information on potential overlaps of all the fragments is generated and the direction of the potential overlap (forward-forward or forward-complement) is generated.

Systematic Match Inspection

In the next fundamental step, potential overlaps found during the scanning phase are examined with a Smith–Waterman based algorithm for local alignment of overlaps. SW alignment algorithm takes into account the fact that today's base caller has a low error rate. The block indel model does not apply to the alignment of shotgun sequencing data. This means that long stretches of mismatches or gaps in the alignment are less probable than small, errors. We also assume an "mismatch" of a base against a 'N' (symbol for a Ny base) to have no penalty, as in many cases the base caller rightfully set 'N' for a real existing base (and not an erroneous extra one) that could not be resolved further.

Quality criteria like the score of the expected length of the overlap, the overall computed score of the overlap, etc. are calculated for each overlap. Every candidate pair, whose computed score is within a configurable threshold of the expected score, and where the length of the overlap is not too small is accepted as "true" overlap, candidate pairs not matching these criteria often due to spurious hits in the scanning phase are identified and rejected from further assembly. The Aligned Dual Sequences (ADS) along with complementary data (like orientation of the aligned reads, overlap region, etc.) that passed the Smith–Waterman test are stored to facilitate and speed up the next phases. Good alternatives are also stored to enable alternative alignments to be found later on in the assembly.

All the ADS form one or several weighted graphs which represent the totality of all the assembly layout possibilities of a given set of shotgun sequencing data. The nodes of the graph are represented by the reads. An edge between two nodes indicates that these two reads are partially overlapping. The weights of the edges themselves are computed from the squared quality of the alignment multiplied by the length of the overlap. This emphasizes the quality aspect of the alignment and also takes the length of the overlap into account.

Building Contigs

The overlaps found and verified in the previous phases must then be assembled into contigs. This is the most fundamental and intricate part of the process, especially in projects containing many repetitive elements. Several basic approaches to the multiple alignment problems have been devised to tackle this problem. Although algorithms for aligning multiple sequences at once have been used with increasing success, lately, for up to about 10 to 15 sequences, the time needed to perform this alignment is still too unpredictable to be used in sequence assembly.

Researchers decided to use iterative pairwise sequence alignment and devise new methods for searching overlap candidates and for empowering contigs to accept or reject reads presented to them during the contig building. The algorithm consists mainly of two objects that interact with each other, a Pathfinder module and a contig building module.

Pathfinder and Contig Interaction

Using an iterative approach to align the multiple alignments is to always successively align an existing consensus against the next read. The results of the alignment sensitivity depend on the order of pairwise alignments. We have to make sure that we start at the position in the contig, where we have many reads with almost no errors. The Pathfinder will thus in the beginning search for a node in the weighted graph having a maximum number of highly weighted edges to neighbours. The idea behind this behaviour is to take the read with the longest and qualitatively best overlaps with as many other reads as possible. This ensures a good starting point, the "anchor" for this contig.

They first tried a simple greedy algorithm to determine the next overlap candidate to a contig, but on occasions especially in highly repetitive parts of a genome, this algorithm fails. With their current algorithm, the number of misalignments could be substantially reduced, the Pathfinder designates the next read to add to an existing contig by making an in-depth analysis 7 of the weights to neighbouring reads, taking the edge leading to the first node that is contained in the best partial path found so far. It then presents this read (and its approximate position) to the contig object as potential candidate for inclusion into the existing consensus.

A contig (contig is a computer program through which we can cut or add a sequence) is represented by a collection of reads that have been arranged in a certain order with given offsets to form an alignment that is as optimal as possible, i.e., an alignment where the reads forming it have as few unexplained errors as possible but still form the shortest possible alignment. To serve this purpose, a contig object has been provided with functions that analyse the impact of every newly added read (at a given position) on the existing consensus. The assumption is now that as the assembly started with the best overlapping reads available, the bases in the consensus will be right at almost every position. Should the newly added read integrate nicely into the consensus, perhaps extending it, and then the contig object will accept this read as part of the consensus. In cases representing a chance alignment, the read differs in too many places from the actual consensus (thus differing in many aspects from reads that have been introduced to the consensus before). The contig will then reject the read from its consensus and tell the Pathfinder object to search for alternative paths through the weighted overlap graph. The Pathfinder object will eventually try to add the same read to the same contig but at a different position or by skipping it or trying other reads.

Once the Pathfinder has no possibilities left to add unused reads to the actual contig, it will again search for a new anchor point and use this as the starting point for a new contig. This loop continues until all the reads have been put into contigs.

Consensus Approval Methods

As mentioned briefly above, each contig object has the possibility to reject a read being introduced into its consensus. This is one of the most crucial points in the development of assembler, allowing presumably good building blocks, i.e., reads with high quality. To start an assembly is a decisive step in the ongoing assembly process. It allows using implicit and explicit knowledge available in reads that are known to be good.

The simplest behaviour of a contig could be to simply accept every new read that is being presented as part of the consensus without further checks. In this case, the assembler would thus rely only on the weighted graph, the method with which the weighted edges were calculated and the algorithm which traverses this graph.

However, using additional information that is available at the time of assembly might prove useful. For example, known standard repetitive elements have been tagged in the data-preprocessing step of the assembly. It is therefore possible to apply much stricter control mechanisms in those stretches of a sequence known to be repetitive, because they might really differ but only in a few bases and are thus dangerous in the assembly.

In a new assembly, repetitive elements must be checked because they will induce errors. Reads with errors will be rejected from the already existing contigs as the Pathfinder tries to enter them. In an automatic assembly process contig object failed to find a valid explanation to this problem when calling the signal analysis function provided by the automatic editor. Automatic editor will resolve the uncertainties by investigating probable alternatives at the fault site.

Highly trained personnel normally execute this labour-intensive task, although in a significant number of cases it is fairly easy to adjudicate between conflicting readings by analysing the trace data. A previous suggestion on incorporating electrophoresis data into the assembly process promoted the idea of cap turning intensity and characteristic trace shape information and provides these as additional data to the assembly algorithm. They decided against such an approach as essential. The signal analysis reveals enough evidence for resolving a conflict within reads by changing the bases incriminated. Signal analysis is therefore treated as a black box decision formed by an expert system called only during the assembly algorithm when conflicts arise. It provides nevertheless more information than the one contained in quality values extracted from the signal of one read only. There is a reasonable suspicion deduced from other aligned reads in that place and the type of error where the base caller could have made a mistake.

There is not enough evidence found in the conflicting reads to allow a base change to resolve the discrepancies arising. There is also the additional knowledge that these non-resolvable errors occur in an area tagged as "*Alu* repeats", which leads to a highly sensitive assessment of the errors. Consequently, the reads 4 and 5 containing repetitive sequence which the first contig object could not match with its consensus, form a second contig.

Read Extension

As the initial assembly uses only high-quality parts of the reads, further information can be extracted from the assembly by examining the end of the reads that were previously unused because the quality seems too low. Although the signal to noise in read traces quickly degrades towards the end, the data is not generally useless. These 'hidden' parts of the reads can now be uncovered in two ways: (i) by uncovering parts of the reads that align to the already existing consensus and (ii) by uncovering hidden stretches of reads at the end of the contigs that are not confirmed by a consensus.

Intracontig Extension is used to uncover reads and "beef up" areas of low coverage within a contig and is a straightforward process. It is mainly used as a method to get more data confirmation than is available. By using only high-quality parts, the hidden sequence is aligned step by step to the existing consensus. In most cases, the discrepancies found between the HCRs forming the existing consensus and the unaligned LCRs will be decided in favour of the HCR. In some cases, especially in regions with very low coverage, one or more reads with LCR data can correct an error in the HCR stretch, e.g. when there is a local drop in the confidence values and signal quality of bases in the HCR stretch, whereas signal quality and confidence values of the same bases in the LCR stretch seem better.

Extracontig Read extension, the second possibility, uncovers LCR at the ends of contigs and is used to extend the consensus to the left or to the right of a contig. LCR data present at the end of probably each read is not of bad quality, but it is treated as hidden data (a region where the base caller has calculated lower quality for the bases because it depended on the trace data of a single read). However, once reads have been aligned in their HCR, two or more stretches of lower quality can be used to uncover each other. The main purpose of this is to enable potential joints between contigs.

The iterative enlargement procedure enables the assembler to redefine the HCR of each read step, by step by comparing it with supporting sequences from aligned reads. This use of information in collateral reads is the assembler's major advantage over a simple base caller, which has only the trace information of one read to call bases.

Contig Linking and Editing

In the last step of the assembly, the extended contigs are linked together where possible and the result is passed to the automatic finisher to correct errors in the assembly. Depending on the number and type of errors found in the assembled data by the finisher, the contigs can be dismantled and reassembled, taking into account the corrections made to the individual reads by the autofinisher and therefore refining the overall assembly.

By cycling through the previous steps, the assembler iteratively corrects errors like misassembled repetitive repeats that were made during previous steps and thus ensures that the resulting contigs contain as few unexplainable errors as possible.

GENOME SEQUENCING METHODS

There are three basic methods to sequence DNA. They are:

- I. Maxam–Gilbert Chemical DNA sequencing
- II. Sanger–Coulson or dideoxy DNA sequencing
- III. Automated DNA sequencing with fluorescent labels

The sequences may be assembled by any one of the following methodology:

a. Shotgun sequencing
b. Clone contig assembly
c. Directed shotgun assembly

Maxam–Gilbert or Chemical Degradation Sequencing *in vitro* DNA Polymerase Reaction

1. Use radioactive labels at one end only (Figure 2.1).
2. In four separate reactions, treat the DNA with base-specific chemicals that result in cleavage of the DNA strand at that base. Example: DMS (dimethyl sulphate) for G's, hydrazine for pyrimidines.

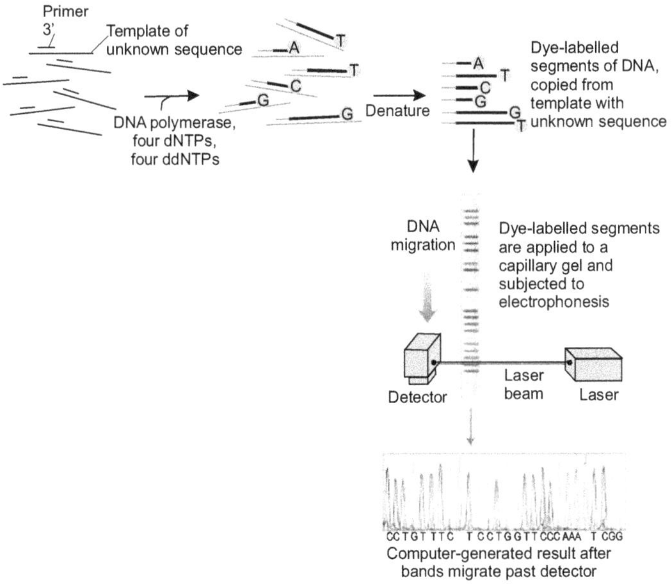

Figure 2.1 Maxam–Gilbert Method

3. Get a nested set of labelled DNA fragments.
4. Analyse as a ladder on a DNA sequencing gel:
   a. Polyacrylamide denaturing gel resolution of short ssDNA fragments
   b. Denaturing gel: 8 M urea. Keep DNA denatured during electrophoresis
   c. Analyse only the labelled DNA fragments via autoradiography or fluorescence analysis
5. Read the DNA sequence from the bottom of the gel to the top by examining the ladder of DNA bands which gives sequence 5′ to 3′ (5′ → 3′)

Sanger–Coulson or Dideoxy DNA Sequencing

1. Comes from DNA polymerase properties: Have a DNA template and a DNA primer (Figure 2.2).

   Cloning vehicles often have universal primers for sequencing DNA cloned into one of the MCS (Multiple Cloning Sites) sites or polylinkers.
2. Execute 4 separate polymerization reactions containing each of the 4 dNTPs and one each of the four dideoxynucleoside TPs: ddGTP, ddATP, ddCTP, ddTTP.

   To assay the product DNA, one of the 4 dNTPs is radioactively labelled, or the primer is labelled, radioactively or using fluorescent dyes, or the ddNTP is labelled with fluorescent dyes. Dideoxy means 3′-H as well as 2′-H. When a ddNTP is incorporated, it acts as a chain terminator: DNA synthesis stops since the DNA primer no longer has a 3′-OH primer terminus.

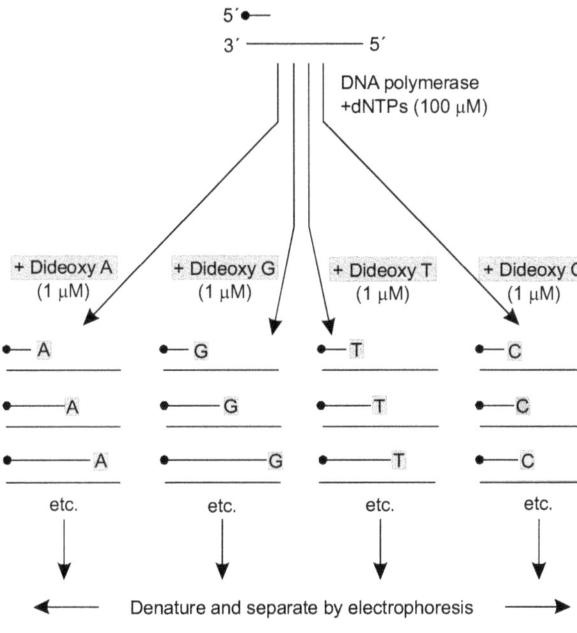

Figure 2.2 Sanger–Coulson method

3. Thus, we get as reaction products, a nested set of fragments, each terminated at one of the four bases: G in the ddGTP reaction, A in the ddATP reaction, and so on.
4. When run on a DNA sequencing gel (polyacrylamide, with DNA denatured), the "nested set" of fragments forms a ladder of DNA bands corresponding to the positions of the bases.
5. Read the DNA sequence by reading the four lanes, one for each base, from bottom-up, to correspond to $5' \rightarrow 3'$ sequence.

Automated DNA Sequencing with Fluorescent Labels

This is the Sanger dideoxy sequencing method but with either fluorescent primers or fluorescent dideoxy chain terminators. Different fluors are used for each of the four nucleotide types. This permits analysis of all four nucleotide reactions for a given DNA sample in one lane of the sequencing gel, approximately fourfold increase in analysis capability per gel run (Figure 2.3).

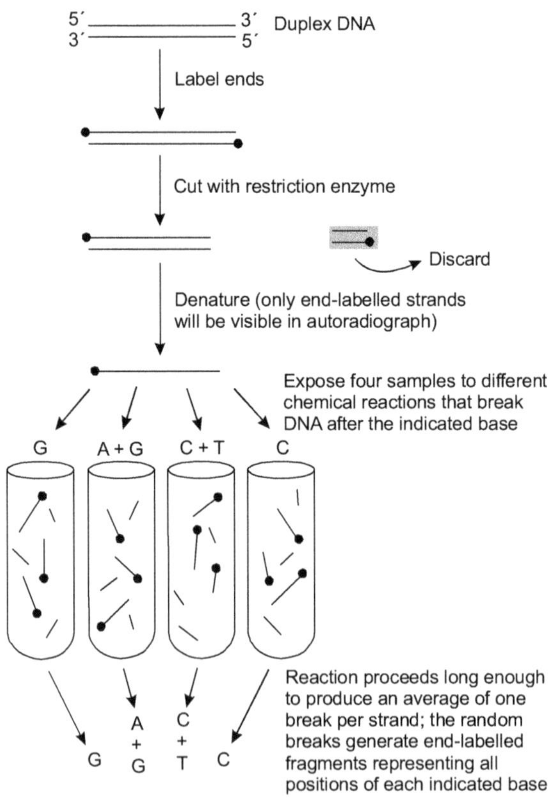

Figure 2.3 Automated sequencing

The most recent capillary gel electrophoresis is used in the Perkin Elmer-AB1 3700 automated sequencing machines rather than slab gel electrophoresis. They have separate thin capillary gel for each DNA sample.

Advantages

1. Better resolution, no running over from one lane to another
2. Separation of bands occurs much faster, i.e., 10 to 15 fold increase in speed

Both advantages are very important for Celera sequencing of the human genome.

GENOME SEQUENCING PROBLEMS

1. The major problem in DNA sequencing is that it can only Sequence 500–700 nucleotides from a given DNA sample. This is due to convergence of the DNA bands. That is, the per cent size difference between bands of 9 and 10 nucleotides is 10% but this percent size difference between bands of 99 and 100 nucleotides is only 1%. Thus, bands corresponding to 99 and 100 nucleotides are 10-fold closer to each other than the bands corresponding to DNA fragments of length 9 and 10 nucleotides. Thus, in genome sequencing or sequencing of DNA molecules much longer than 500–700 nucleotides, one must obtain sequences of many overlapping sets of ~ 500 bp fragments and then

join these together by determining how they overlap each other.

2. During DNA sequence assembly overlapping sequences are joined together to form contigs. When overlapping sequences are properly joined to form a single sequence, this single sequence is called a contig. In sequence assembly for an entire genome, ultimately one should end up with a single contig for each chromosome, since each chromosome is composed of a single DNA molecule. In practice, this is very difficult, due to repetitive DNA sequences.

Repetitive DNA sequences present two major problems for sequence assembly:

1. ***Number of repeat copies*** If the length of a repetitive DNA sequence region is long compared to the sequenced DNA length of ~500 bp, it is nearly impossible to determine how many copies of the DNA repeat are present.

2. ***Correct assembly*** If such a long repetitive DNA sequence region is present at several sites on a genome, then it is nearly impossible to determine what DNA sequences should be properly joined on either side of each repetitive DNA sequence region.

For these reasons (and a few others), in genome sequencing of genomes from higher eukaryotes, sequence is not obtained for much repetitive DNA and a given chromosome sequence will be present in several contigs in the final assembly.

METHODOLOGY FOR DNA SEQUENCE ASSEMBLY

There are three main methodology (with variations on a theme):

1. Shotgun sequencing
2. Clone-contig assembly
3. Directed shotgun assembly

Shotgun Sequencing

One obtains a high redundancy of sequencing of a given long DNA: 10–15 fold. One then uses computer programs to find the correct overlaps and join individual sequence reads into long contigs. To do this uniquely and correctly, one needs:

1. Little repetitive DNA and
2. Overlaps between reads of 20–40 nucleotides. Such overlaps necessitate the high degree of redundancy.

Closure of gaps One will still have some gaps that need closing. This is done via:

1. *Use of a second clone library*—often using a different cloning vehicle. This will often yield a clone which will cover the gap.
2. *Use of directed sequencing*—from this clone, obtain initial ~500 bp of sequence from one end. Then use this sequence to construct an oligonucleotide

to use as primer to extend the sequence further into the clone, internal primer.

Continue doing this until one has walked across the gap, thereby closing the gap.

Clone-contig Assembly

First generate a collection of mapped clone fragments:

1. **Examples** YACs, BACs, PACs, cosmids which are mapped relative to each other, forming a set of overlapping large-cloned fragments often with a high degree of redundancy 5–15 fold.
2. From among these, a minimum set of overlapping clones are chosen. This is sometimes called a minimum tiling clone set.
3. For each of these overlapping clones, shotgun sequencing was done. Re-clone or subcloning each large overlapping clone as small sequencing fragments, is done. These sequence ultimately forms a single contig corresponding to the sequence of each large clone.
4. Each contig for each large clone is joined together via the overlapping sequence and knowledge of the map of the clones, yielding the final sequence of the entire DNA molecule, e.g. a chromosome.

How is the collection of mapped clone fragments generated?

1. Generate a genome library as YACs, BACs, PACs, cosmids, etc.

(a) Locate markers to specific clones in the library. These markers can be genetic markers, e.g. genes, or physical DNA markers, e.g. Rsites, STSs, RFLPs, etc.

(b) Identify overlapping clones by identifying pairs of clones uniquely containing the same markers, e.g. genes or shared R fragments.

2. When these markers are STSs, they can serve as physical anchors in the sequence assembly process, i.e., one knows the position of a specific sequence from the position of the STS and the sequence assembly must have this sequence in this position to be correct assembly.

Chromosome walking One can also determine overlapping clones without specific markers by hybridizing one clone DNA to the DNA of other clones.

However, use of STSs to provide anchors are very desirable with very large DNA fragments.

Directed Shotgun Assembly

This is shotgun DNA sequencing and assembly of very large genomes, e.g. *Drosophila* or human coupled with use of anchored DNA markers, e.g. STSs.

Example Celera approach to sequencing the human genome. Three genomic DNA libraries are used for sequencing:

1. Plasmid library with ~2 kb inserts

2. Library with ~10 kb inserts different cloning vehicle used 10 kb is large compared with most repetitive DNA regions in the human genome, thereby avoiding much of this problem
3. BAC clone library with ~250 kb inserts.

DNA sequencing is done on both ends of the inserts present in each of these clones.

Computerized assembly of sequence into contigs is greatly helped by the fact that the distance between the sequences of each of the two ends is nearly constant, at 2 kb or 10 kb or 250 kb: these are called "sequence pairs".

The sequences from the BAC clone ends are used in two ways:

1. They become STSs
2. These STSs are used to map the BACs and the STSs
3. The mapped STSs become anchored sequences for the subsequence assembly into contigs

Note the difference here from clone-contig assembly:

1. In clone-contig assembly, mapping is done first, a minimum set of YACs or BACs is determined and then each of these YACs or BACs is subjected to shotgun sequencing.
2. Here, in directed shotgun assembly, the entire genome is subjected to Shotgun sequencing via the two smaller insert libraries, the 2 kb and 10 kb insert libraries.

These methodologies yield then the ultimate physical map of the genome.

This DNA sequence is the foundation basic about the organism. With this information, one completely knows the enzymatic and molecular capabilities of the organism.

OTHER METHODS OF DNA SEQUENCING

"Sequencing by hybridization" is a non-enzymatic method that uses a DNA microarray. In this method, a single pool of unknown DNA is fluorescently labelled and hybridized to an array of known sequences. If the unknown DNA hybridizes strongly to a given spot on the array, causing it to 'light up', then that sequence is inferred to exist within the unknown DNA being sequenced.[7]

Mass spectrometry can also be used to sequence DNA molecules; conventional chain-termination reactions produce DNA molecules of different lengths and the length of these fragments is then determined by the mass differences between them (rather than using gel separation).[8]

Other methods include labelling the DNA polymerase, reading the sequence as a DNA strand transits through nanopores, and microscopy-based techniques, such as AFM or electron microscopy that are used to identify the positions of individual nucleotides within long DNA fragments (>5,000 bp) by nucleotide labelling with heavier elements (e.g. halogens) for visual detection and recording.

IMPORTANT DNA SEQUENCE CHRONICLE

1953 Discovery of the structure of the DNA double helix

1972 Development of recombinant DNA technology

1974 EMBL-bank, the first nucleotide sequence repository, is started at the European Molecular Biology Laboratory

1975 The first complete DNA genome to be sequenced is that of bacteriophage ϕX174

1977 Allan Maxam and Walter Gilbert—DNA sequencing by chemical degradation[9]

1977 Fred Sanger—DNA sequencing by enzymatic synthesis

1980 Fred Sanger and Wally Gilbert receive the Nobel Prize in Chemistry

1982 Genbank starts as a public repository of DNA sequences

1982 Andre Marion and Sam Eletr from Hewlett Packard start Applied Biosystems for automated sequencing

1982 Akiyoshi Wada proposes automated sequencing and gets support to build robots with help from Hitachi

1984 Medical Research Council scientists decipher the complete DNA sequence of the 170-kb Epstein–Barr virus

1985 Kary Mullis and colleagues develop the polymerase chain reaction (PCR)

1986 Leroy E. Hood's laboratory at the California Institute of Technology and Smith announce the first semi-automated DNA sequencing machine

1987 Applied Biosystems markets first automated sequencing machine (model ABI 370)

1987 Walter Gilbert starts Genome Corp., with the goal of sequencing and commercializing the data

1990 The U.S. National Institutes of Health (NIH) begins large-scale sequencing trials at 75 cents (US)/base

1990 Barry Karger, Lloyd Smith and Norman Dovichi—capillary electrophoresis[10, 11 & 12]

1990 Craig Venter develops strategy to find expressed genes with ESTs.

1991 Uberbacher develops GRAIL, a gene-prediction program

1992 Craig Venter set up The Institute for Genomic Research

1992 William Haseltine heads Human Genome Sciences, to commercialize TIGR products

1992 Wellcome Trust begins participation in the Human Genome Project

1992 Simon *et al.* develop BACs (Bacterial Artificial Chromosomes) for cloning

First chromosome physical maps published:

Page *et al.*—Y chromosome[13]

Cohen *et al.*—chromosome 2

Lander—complete mouse genetic map[14]

Weissenbach—complete human genetic map[15]

1993 Wellcome Trust and MRC open Sanger Centre, near Cambridge, UK

1993 The Genbank database migrates from Los Alamos (DOE) to NCBI (NIH)

1995 Venter, Fraser and Smith publish first sequence of free-living organism, *Haemophilus influenzae*

1995 Richard Mathies *et al.* publish on sequencing dyes[16]

1995 Michael Reeve and Carl Fuller, thermostable polymerase for sequencing[17]

1996 International HGP partners agree to release sequence data into public databases within 24 hours

1996 International consortium releases genome sequence of yeast *S. cerevisiae*

1996 Yoshihide Hayashizaki's at RIKEN completes the first set of full-length mouse cDNAs

1996 ABI introduces a capillary electrophoresis system, the ABI 310 sequence analyser

1997 Blattner, Plunkett *et al.* publish the sequence of *E. coli*[18]

1998 Phil Green and Brent Ewing of Washington University publish "Phred" for interpreting sequencer data[19]

1998 Venter starts new company "Celera"

1998 Applied Biosystems introduces the 3700 capillary sequencing machine

1998 Wellcome Trust doubles support for the HGP to $330 million for 1/3 of the sequencing

1998 NIH and DoE goal: "working draft" of the human genome

1998 Sulston, Waterston *et al.* finish sequence of *C. elegans*[20]

1999 NIH moves up completion date for rough draft to 2000

1999 NIH launches the mouse genome sequencing project

1999 First sequence of human chromosome 22 published

2000 Celera and collaborators sequence *Drosophila melanogaster*

2000 Validation of Venter's shotgun method

2000 HGP consortium publishes sequence of chromosome 21

2000 HGP and Celera jointly announce working drafts of HG sequence

2000 Estimates for the number of genes in the human genome range from 35,000 to 120,000

2000 International consortium completes first plant sequence, *Arabidopsis thaliana*

2001 HGP consortium publishes Human Genome Sequence draft in Nature[21]

2001 Celera publishes the human genome sequence[22]

2005 420,000 VariantSEQr human resequencing primer sequences published on new NCBI probe database

2006 The X prize foundation established the Archon X prize, intending to award $10 million to the first team that can build a device and use it to sequence 100 human genomes within 10 days or less, with an accuracy of no more than one error in every 100,000 bases sequenced, with sequences accurately covering at least 98% of the genome, and at a recurring cost of no more than $10,000 (US) per genome

2007 For the first time, a set of closely related species (12 Drosophilidae) are sequenced, launching the era of phylogenomics

2007 Craig Venter publishes his full diploid genome—the first human genome to be sequenced completely

2008 An international consortium launches the 1000 genomes project, aimed to study human genetic variability

2008 Leiden University Medical Center scientists decipher the first complete DNA sequence of a woman.

REVIEW QUESTIONS

1. Describe the Maxam–Gilbert method of DNA sequencing.
2. Elaborate the Sanger–Coulson method of DNA sequencing.
3. Write short notes on automated sequencing method. Add a note on the advantages and disadvantages of the same.

REFERENCES

1. Sanger, F. and Nicklen, S. Coulson. (1977). DNA sequencing with chain-terminating inhibitors. Proceedings of the National Academy of Sciences. *Natn. Acad. Sci.* USA. 74: 5463–5467.
2. Dear, S., Durbin, R., Hillier, L., Marth, G., Thierry-Mieg, J. and Mott, R. (1998). "Sequence assembly with CAFTOOLS." *Genome Res.* 8: 260–267.
3. Hannu Peltola, Hans Söderlund and Esko Ukkonen. (1984). "SEQAID: A DNA sequence assembling program based on a mathematical model." *Nucleic Acids Research.* 12(1): 307–321.
4. Parsons, R., Forrest, S. and Burks, C. (1993). "Genetics algorithms for DNA sequence assembly." *Proc. Int. Conf. Intell. Syst. Mol. Biol.* 1: 310–8.
5. Notredame, C. and Higgins, D.G. (1996). "Saga-sequence alignment by genetic algorithm." *Nucl. Acid. Res.* 24 (8): 1515–1524.
6. Ching Zhang and Andrew, K.C. Wong. (1997). "A genetic algorithm for multiple molecular sequence alignment." *Bioinformatics*. Volume 13, Number 6 pp. 565–581.
7. Hanna, G.J., Johnson, V.A., Kuritzkes, D.R., Richman, D.D., Martinez-Picado, J., Sutton, L., Hazelwood, J.D. and D'Aquila, R.T. (2000). "Comparison of sequencing by hybridization and cycle sequencing for genotyping of human immunodeficiency virus type 1 reverse transcriptase." *Journal of Clinical Microbiology.* 38(7): 2715.

8. Edwards, J.R., Ruparel, H. and Ju, J. "Mass-spectrometry DNA sequencing." *Mutation Research.* 573 (1–2): 3–12.

9. Maxam, A.M. and Gilbert, W. (1977). "A new method for sequencing DNA." *Proc. Natl. Acad. Sci.* USA. Feb. 74(2): 560–4.

10. Karger, L. Barry, Guttman, A., Cohen, A.S. and Heiger, D.N. (1990). "Analytical and micropreparative ultra-high resolution of oligonucleotides by polyacrylamide gel high-performance capillary electrophoresis." *Analytical Chemistry.* 62 (2): 137–141.

11. Smith, M. Lloyd, Luckey, J.A., Drossman, H., Kostichka, A.J., Mead, D.A., D'Cunha, J. and Norris, T.B. (1990). "High speed DNA sequencing by capillary electrophoresis." *Nucleic Acids Research.* 18: 4417–4421.

12. Dovichi, J. Norman, Swerdlow, H.P., Wu, S. and Harke, H.R. (1990). "Capillary gel electrophoresis for DNA sequencing: laser-induced fluorescence detection with the sheath flow cuvette." *Journal of Chromatography.* 516: 61–67.

13. Page, D.C., Foote, S., Vollrath, D.and Hilton, A. (1992). "The human Y chromosome: overlapping DNA clones spanning the euchromatic region." *Science.* 258 (5079): 60–66.

14. Lander, E.S., Dietrich, W., Katz, H., Lincoln, S.F., Shin, H.S., Friedman, J. and Dracopoli, N.C. (1992). "A genetic map of the mouse suitable for typing intraspecific crosses." *Genetics.* 131: 423–447.

15. Weissenbach, Jean, Gyapay, G., Dib, C., Vignal, A., Morissette, J., Millasseau, P., Vaysseix, G. and Lathrop,

M. (1992). "A second-generation linkage map of the human genome." *Nature.* 359: 794–801.

16. Mathies, R.A., Ju, J., Ruan, C., Fuller, C.W. and Glazer, A.N. (1995). "Fluorescence energy transfer dye-labeled primers for DNA sequencing and analysis." *PNAS* 92: 4347–4351.

17. Reeve, A. Michael, and Fuller, W. Carl. (1995). "A novel thermostable polymerase for DNA sequencing." *Nature.* 376 (6543): 796–797.

18. Blattner, F.R., Frederick, R. Blattner, Guy Plunkett III, Craig, A. Bloch, Nicole, T. Perna, Valerie Burland, Monica Riley, Julio Collado-Vides, Jeremy, D. Glasner, Christopher, K. Rode, George, F. Mayhew, Jason Gregor, Nelson Wayne Davis, Heather, A. Kirkpatrick, Michael, A. Goeden, Debra, J. Rose, Bob Mau and Ying Shao (1997). "The complete genome sequence of *Escherichia coli* K-12." *Science.* 277 (5331): 1453–1462.

19. Ewing, B. and Green, P. (1998). "Base-calling of automated sequencer traces using phred. II. Error probabilities." *Genome Research.* 8 (3): 186–94.

20. The *C. elegans* Sequencing Consortium. (1998). "Genome sequence of the nematode *C. elegans*: A platform for investigating biology." *Science.* 282 (5396): 2012–2018.

21. Lander, and Eric S. (2001). "Initial sequencing and analysis of the human genome." *Nature.* 409 (6822): 860–921.

22. Venter, J.C. *et al.* (2001). "The sequence of the human genome." *Science.* 291 (5507): 1304–51.

3

GENOME MAPPING

This is the first chromosome (21st chromosome) to be mapped.

Daniel Cohen

(On mapping the first chromosome at the French Centre for Genetic Research)

OBJECTIVES

1. To understand the positions of various genes on any genome
2. To get a clear picture of the various types of genome maps

INTRODUCTION

Genome maps serve to guide a scientist toward a gene, just like an interstate map guides a driver from city to city. Like road maps, a genome map is a set of landmarks that tells people where they are. The landmarks on a genome map might include short DNA sequences, regulatory sites that turn genes on and off, and genes themselves. Often, genome maps are used to help scientists find new genes.

The goal of genomics is to determine the position and function of every gene in the genome. This task began in the early 1900s by constructing "maps", in an attempt to identify the positions of loci responsible for particular phenotypes and later, the mutation responsible for a specific variant. The initial steps in mapping are:

1. to establish the proximity of genes or traits to one another and
2. to assign the genes to a particular chromosome.

APPEARANCE OF A GENOME MAP

A genome map is one-dimensional—it is linear like the DNA molecules that make up the genome itself. A genome map looks like a straight line with landmarks noted at irregular intervals along it, much like the towns along the map of a highway. The landmarks are usually inscrutable combinations of letters and numbers that stand for genes or other features—for example, D14S72, GATA-P7042, and so on.

A GENOME MAP AND A GENOME SEQUENCE—DIFFERENT?

Both are portraits of a genome, but a genome map is less detailed than a genome sequence. A sequence spells out the order of every DNA base in the genome, while a map simply identifies a series of landmarks in the genome.

Sometimes mapping and sequencing are completely separate processes. For example, it is possible to determine the location of a gene to "map" the gene without sequencing it. A map may say nothing about the sequence of the genome and a sequence may say nothing about the map.

In other cases, the landmarks on a map are DNA sequences, and mapping is the cousin of sequencing.

MAP TYPES

The two broad categories of maps are genetic maps and physical maps. Both genetic and physical maps provide the likely order of items along a chromosome. Physical maps are more similar to street maps and allow a scientist to more easily home in on a gene's location.

Physical Maps

Physical maps can be divided into three general types: chromosomal or cytogenetic maps, radiation-hybrid (RH) maps and sequence maps. The different types of maps vary in their degree of resolution, that is, the ability to measure the separation of elements that are close together.

Chromosomal or Cytogenetic Map

The lowest-resolution physical map is the chromosomal or cytogenetic map, which is based on the distinctive banding patterns observed by light microscopy of stained chromosomes. Compared with genetic linkage mapping, chromosomal mapping can be used to locate genetic markers defined by traits observable only in whole organisms. Because chromosomal maps are based on estimates of physical distance, they are considered to be physical maps. Yet, the number of base pairs within a band can be estimated.

Radiation-Hybrid Map

RH maps on the other hand, are more detailed. RH maps are similar to linkage maps in that they show estimates of distance between genetic and physical markers, but that is where the similarity ends. RH maps are able to provide more precise information regarding the distance between markers.

Sequence Map

The physical map that provides the most detail is the sequence map. Sequence maps show genetic markers, as well as the sequence between the markers, measured in base pairs.

Genetic Maps

Genetic-linkage or meiotic maps illustrate the order of genes on a chromosome and the relative distances between those genes. Originally, these maps were made by tracing the inheritance of multiple traits, such as hair colour and eye colour, through several generations. Linkage maps can tell where the markers are in relation to each other on the chromosome, but the actual "mileage" between those markers may not be so well-defined.

Genetic-linkage mapping is possible because of a normal biological process called crossing over, which occurs during meiosis—a type of cell division for making sperm and egg cells. During one stage of meiosis, chromosomes line up in pairs along the centre of a cell, where they sometimes "stick" to each other and exchange equivalent pieces of themselves. This sticking and exchanging is called crossing over, and is a relatively common event: on average, a chromosome pair undergoes crossing over about 1.5 times during the formation of each sex cell in humans.

A.H. Sturtevant, then a student at Columbia University, made the first genetic-linkage map of fruit fly genes in 1913—decades before scientists even knew that genes are made of DNA. He found, for example, that leg length was inherited with eye colour more often than with wing length, and that wing length was inherited with eye colour more often than with leg length. Thus, he concluded that the gene for eye colour must be between the genes for wing length and leg length in the fruit fly genome.

CREATING A GENETIC MAP

To produce a genetic map, researchers collect blood or tissue samples from family members where a certain disease or trait is prevalent. Using various laboratory techniques, the scientists isolate DNA from these samples and examine it for the unique patterns of bases seen only in family members who have the disease or trait. These characteristic molecular patterns are referred to as polymorphisms, or markers.

Before researchers identify the gene responsible for the disease or trait, DNA markers can tell them roughly where the gene is on the chromosome. This is possible because of a genetic process known as recombination. As eggs or sperms develop within a person's body, the 23 pairs of chromosomes within those cells exchange—or recombine—genetic material. If a particular gene is close to a DNA marker, the gene and marker will likely stay together during the recombination process, and be passed on together from parent to child. So, if each family member with a particular disease or trait also inherits a particular DNA marker, chances are high that the gene responsible for the disease lies near that marker.

The more DNA markers there are on a genetic map, the more likely it is that one will be closely linked to a disease gene, and the easier it will be for researchers to zero-in on that gene. One of the first major achievements of the HGP was to develop dense maps of markers spaced evenly across the entire collection of human DNA.

Genetic linkage maps of each chromosome are made by determining how frequently two markers are passed together from parent to child. Because genetic material is sometimes exchanged during the production of sperm and egg cells, groups of traits (or markers) originally together on one chromosome may not be inherited together. Closely linked markers are less likely to be separated by spontaneous chromosome rearrangements. In Figure 3.1, the vertical lines represent chromosome 4 pairs for each individual in a family. The father has two traits that can be detected in any child who inherits them: a short known DNA sequence used as a genetic marker (M) and Huntington's disease (HD). The fact that one child received only a single trait (M) from that particular chromosome indicates that the fathers' genetic material recombined during the process of sperm production. The frequency of this event helps determine the distance between the two DNA sequences on a genetic map.

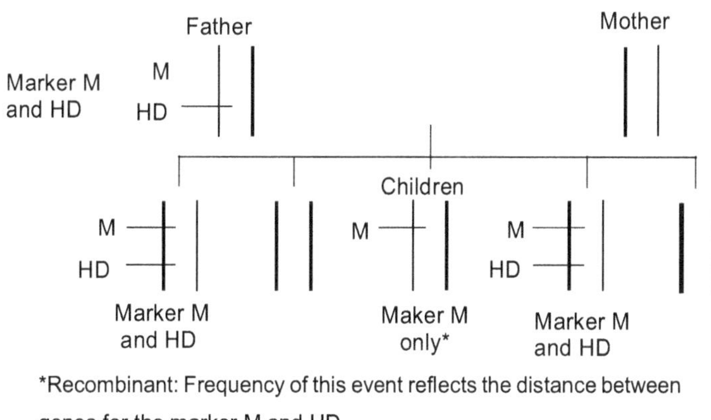

Figure 3.1 Constructing a genetic linkage map

Contig or Clone-Contig Map

This is a genetic map depicting the relative order of a linked (contiguous) library of small overlapping clones representing a complete chromosomal segment. Contig maps provide the possibility to study a complete and often large segment of the genome by examining a series of overlapping clones which then provide an unbroken succession of information about that region. Being a map depicting the relative order of a linked library of small overlapping clones, it represents a complete chromosome segment showing the locations of those regions of a chromosome, where contiguous DNA segments overlap. These maps are important because they provide the ability to study a complete, and often large, segment of the genome by examining a series of overlapping clones which then provide an unbroken succession of information about that region.

DNA MARKERS

Markers themselves usually consist of DNA that does not contain a gene, however they can tell a researcher the identity of the person from whom the DNA sample came. This makes markers extremely valuable for tracking inheritance of traits through generations of a family, and markers have also proven useful in criminal investigations and other forensic applications.

Although there are several different types of genetic markers, the type most used on genetic maps today is known as a **microsatellite map**. However, maps of even higher resolution are being constructed using single-nucleotide polymorphisms, or SNPs (pronounced "snips"). Both types of markers are easy to use with automated laboratory equipment, so researchers can rapidly map a disease or trait in a large number of family members.

The development of high-resolution, easy-to-use genetic maps, coupled with the HGP's successful sequencing and physical mapping of the entire human genome, has revolutionized genetics research. The improved quality of genetic data has reduced the time required to identify a gene from a period of years to, in many cases, a matter of months or even weeks. Genetic mapping data generated by the HGP's laboratories is freely accessible to scientists through NCBI database, e.g. Restriction Fragment Length Polymorphisms (RFLPs), Variable Number of Tandem Repeat Polymorphisms (VNTRs), Microsatellite Polymorphisms, Single Nucleotide Polymorphisms (SNPs), etc.

HAPPY Maps

HAPPY mapping is an *in vitro* approach for defining the order and spacing of DNA markers directly on native genomic DNA. This cloning-free technique is based on analysing the segregation of markers amplified from high-molecular weight genomic DNA which has been broken randomly and "segregated" by limiting dilution into subhaploid samples. It is a uniquely versatile tool, allowing for the construction of genome maps with flexible ranges and resolutions. Moreover, it is applicable to plant genomes, for which many of the techniques pioneered in animal genomes are inapplicable or inappropriate. Such maps can greatly aid the construction of regional or genome-wide physical maps.

GENETIC MAPS AS A FRAMEWORK FOR PHYSICAL MAP CONSTRUCTION

Genetic maps (Figure 3.2) are also used to generate the essential backbone, or scaffold, needed for the creation of more detailed human genome maps. These detailed maps called physical maps, (Figure 3.3), further define the DNA sequence between genetic markers and are essential to the rapid identification of genes.

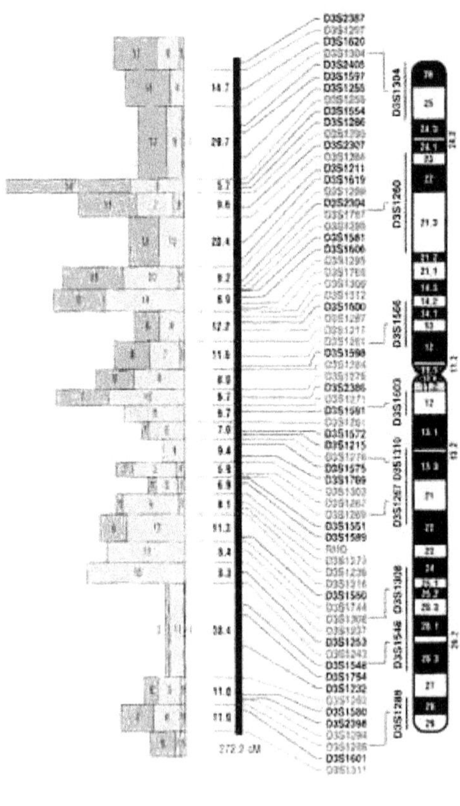

Figure 3.2 Human chromosome 3 (Genetic map). (Source: *Science*)

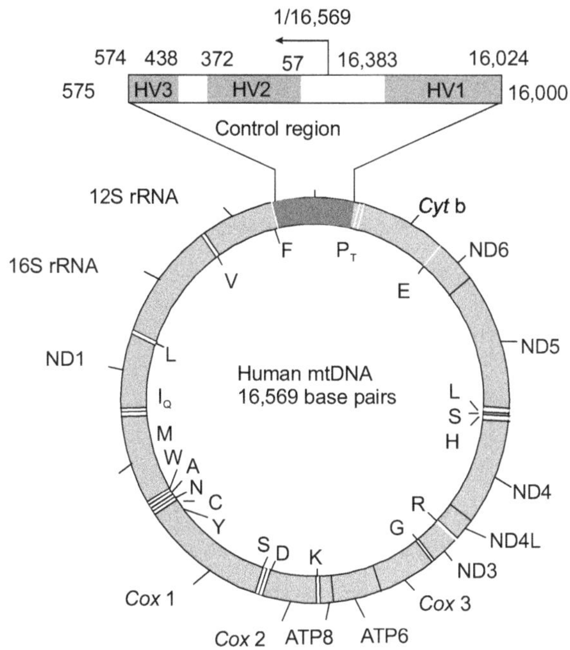

Figure 3.3 Human mtDNA physical map

LIMITATIONS TO CONSTRUCTING A GENETIC MAP

- Access to polymorphic traits or markers
- Need for a large number of progeny and/or multiple generations
- Best performed in model organisms subject to selective breeding
- Crossing-over does not occur at random (maps of limited accuracy)

In contrast, some form of physical map can be constructed for any organism.

MAPPING OF GENOMES—AN OUTLINE

i. Gene isolation—reverse genetics
ii. Genome cloning—genome libraries
iii. Mapping from cloning—sequence tagged sites (STSs)
iv. cDNA cloning—expressed sequence tags (ESTs)
v. Radiation hybrids
vi. Mapping from radiation hybrids: STSs

RESEARCH SCOPE

A classical example of research potential in this field is the announcement of an international research consortium's 1000 Genomes Project. It is an ambitious effort that will involve sequencing the genomes of at least a thousand people from around the world to create the most detailed and medically useful map to date of human genetic variation. The project will receive major support from the Wellcome Trust Sanger Institute in Hinxton, England, the Beijing Genomics Institute, Shenzhen (BGI Shenzhen) in China and the National Human Genome Research Institute (NHGRI), part of the National Institutes of Health (NIH).

USES OF GENOME MAPS

Genome maps help scientists to find genes, particularly those involved in human disease. It helps to study many families affected by a disease, tracing the inheritance of the disease and of specific genome landmarks through several generations. Landmarks that tend to be inherited along with the disease are likely to be located close to the disease gene and become "markers" for the gene in question, e.g. SNP in sickle-cell anaemia.

Once a few such markers have been identified, the approximate location of the disease gene is known. In this way, narrowing down the search from an entire 3-billion-base-pair genome to a region of the genome a few million base pairs long is accomplished. The next step is to look for genes in that part of the genome and study the genes one by one, to learn which one is involved in the disease.

Genes for cystic fibrosis, Huntington's disease and many other inherited diseases have been identified by this method, but it is a time-consuming and laborious process. Several million base pairs are still a lot of DNA and a region of the genome that size may contain dozens of genes for scientists to sort through. Imagine you have a road map that can get you to the right neighbourhood, but then you have to drive around and knock-on doors until you find the house you're looking for.

In addition to helping in the search for genes, genome maps are useful in the day-to-day activities of molecular biology laboratories. In the lab, the human genome lives in the form of clones which are nothing but chunks of DNA that have been chopped up and spliced into the DNA of bacteria or other cells. This method keeps each chunk of the genome separate from the others and available in many copies for easy experiment and study.

Genome maps also help scientists to find and learn about other important parts of the genome, such as the regulatory regions that help control when genes are turned on and off. They illuminate the overall structure of the genome—places where several related genes are clustered together, for example, or parts of the genome that contain an unusually rich concentration of genes. Finally, genome maps enable us to compare the genomes of different species, yielding insight into the process of evolution.

Other Uses

1. **As markers for genetic mapping,** e.g. pedigree analysis.
2. **As markers for physical mapping,** e.g. restriction mapping.
3. **In the study of genetic disease,** to compare diseased vs non-diseased individuals.
4. **For forensic purposes** to compare DNA from blood, from victim and suspect, compare with blood at "scene of the crime".
5. **Identification purposes** to compare DNA from blood at some site with DNA stock. Military, accidents, etc.
6. In **parentage identification.**
7. **As restriction maps** to locate the positions of and distances between endonuclease recognition sites on a DNA molecule.
8. **As long-range restriction maps** to locate the positions of rare-cutting endonuclease recognition sites on a DNA molecule by PFGE.

9. **As clone (contig) maps** that consist of libraries of overlapping clones where the relationship of each clone to other clones has been resolved.
10. **In fluorescent *in situ* hybridization (FISH)** to locate the position of a marker by hybridizing a labelled probe to intact chromosomes.
11. **As optical maps** that visually inspect and measure the positions of endonuclease recognition sites on a DNA molecule.
12. **As EST maps** to plot the location of transcribed sequences.

SOME GENOME MAPPING AND VIEWING TOOLS/SOFTWARE

1. **NCBI's Map Viewer**—a tool for integrating genetic and physical maps
2. **AceDraw**—designed for the organization and management of mapping data and easy map drawing
3. **ALLASS**—to map a disease locus by allelic association
4. **ALOHOMORA**—to facilitate genome-wide linkage studies performed with high-density SNPs
5. **AntMap**—constructing genetic linkage maps
6. **APL-OSA**—subset analysis in genetic linkage mapping of complex traits
7. **BMAPBUILDER**—building chromosome-wide maps
8. **CARTHAGENE**—genetic/radiation hybrid mapping software

9. **CHECKMATRIX**—a visualization tool to validate constructed genetic maps
10. **CHROMSCAN**—a statistical based program for association mapping of disease genes
11. **CMAP**—a web-based tool that allows users to view comparisons of genetic and physical maps
12. **COMBIN**—for the construction of highly saturated linkage maps
13. **CRIMAP**—for constructing multilocus linkage map
14. **DMLE** (Disease Mapping using Linkage disEquilibrium)—high-resolution mapping of the position of a disease mutation relative to a set of genetic markers
15. **T-DNA Express**—*Arabidopsis* gene mapping tool

EMERGING MAPS

The genome's cartographers are now making maps that combine features of both genetic-linkage and physical maps. For example, new methods of genetic-linkage mapping based on polymorphisms enable scientists to place landmarks at the proper physical locations on the chromosomes. Meanwhile, some STSs on physical maps are parts of genes and thus give the sort of information that appears on genetic-linkage maps.

As mapping techniques advance, scientists try to create maps with more landmarks that are more closely, evenly, and accurately spaced. But in contrast to DNA sequencing, which has become increasingly automated, genome mapping still can only be accomplished by experienced scientists. And even the most expert mapper may run into difficulties finding the desired number and type of landmarks with the desired spacing. This means that although maps keep improving, the work is slow and there is still a long way to go.

It is almost as difficult and arbitrary to define a "complete" genome map as it is to define a "complete" genome sequence. One definition of "complete" is a map that includes the sequence and location of all of an organism's genes. Such maps currently exist for more than 150 organisms, most of them viruses with small genomes.

DIFFERENCES BETWEEN PHYSICAL AND GENETIC MAPS

Genetic maps	Physical maps
Collinear with physical placement of genes on DNA	Placement of genes and sites on nucleotide sequence of DNA
Based on genes and mapping via recombination frequency	Based on sites on DNA and mapping via locating the site on the DNA
Genetic maps depict relative positions of loci based on the degree of recombination	Physical maps show the actual (physical) distance between loci
This approach studies the inheritance/assortment of traits by genetic analysis	This approach applies techniques of molecular biology
Many factors limit the construction of these maps	Can be constructed for any organism

REVIEW QUESTIONS

1. Write an account of physical maps.
2. Elaborate on genetic maps.
3. Give an account of the various methods of physical mapping.
4. Distinguish between physical and genetic mapping strategies.
5. What are uses of various types of genome mapping?

REFERENCES

1. http://www.ornl.gov/sci/techresources/Human_Genome/publicat/primer/prim2.html
2. www.genomenewsnetwork.org/resources/whats_a_genome/Chp3_1.shtml
3. www.ncbi.nlm.nih.gov/About/primer/mapping.html
4. www.ornl.gov/TechResources/Human_Genome/publicat/primer/prim2.html
5. http://www.pasteur.fr/recherche/unites/biophyadn/e-Fmapping.html
6. www.broad.mit.edu/cgi-bin/contig/phys_**map**
7. www.wehi.edu.au/MalDB-www/**genome**Info/**Map**Data/**Map**Data.html
8. www.ornl.gov/sci/techresources/Human_**Genome**/research/**mapping**.shtml
9. www.nature.com/nature/journal/v437/n7063/full/nature04226.html

10. bioinformatics.oxfordjournals.org/cgi/content/abstract/24/19/2215
11. www.ncbi.nlm.nih.gov/SCIENCE96/
12. www.ncbi.nlm.nih.gov/gene**map**99/www.molecular-plant-biotechnology.info/molecular-**maps**-of-animal-**genome**/molecular-**maps**-of-animal-**genome**.htm

4

BIOLOGICAL DATABASE MANAGEMENT SYSTEM

Biomedical databases are as much a part of scientists' basic equipment as test tubes and reagents.

Zeeya Merali and Jim Giles

OBJECTIVES

1. To appreciate the approaches to creation, management, analysis and dissemination of biological knowledge
2. To familiarize with the most common and important biological databases

INTRODUCTION

A database is an electronic filing system with a collection of information organized in such a way that a computer program can quickly select desired pieces of data. Traditional databases are organized by fields, records and files. A field is a single piece of information; a record is a complete set of fields; and a file is a collection of records. For example, a telephone book is analogous to a file. It contains a list of records, each of which consists of three fields name, address and telephone number.

To access information from a database, one needs a database management system (DBMS). This is a collection of programs that enable us to enter, organize and select data in a database. Increasingly, the term database is used as shorthand for database management system.

An alternate concept in database design is known as hypertext. Hypertext is a text which is not constrained to be linear and it contains links to other texts. The term was coined by Ted Nelson around 1965. HyperMedia is a term used for hypertext which is not constrained to be text: it can include graphics, video and sound. Hypertext databases are particularly useful for organizing large amounts of information but they are not designed for numerical analysis.

DATA MODELS

The database should be viewed as a representation or model of the world developed for a very specific application. This information will help one organize it in a way that will prove useful. One of the reasons that there are so many software and hardware systems employed for databases, is that each system allows users to represent and model certain types of phenomena.

A data model is an abstract model that describes how data is represented and accessed. They formally define data objects and relationships among data objects. Some applications of database models are development of databases and enabling the exchange of data for a particular area of interest.[1]

Organizing Data

It is important to realize that data can be filled away in several different forms depending on how it needs to be used and accessed. Given below are ways in which data can be stored in databases.

Flat files and Spread sheets Flat files or spread sheets are methods for storing data. All records in this database have the same number of fields. Individual records have different data in each field with one field serving as a key to locate a particular record. When the number of fields becomes lengthy, a flat file is cumbersome to search. The key field is usually determined by the programmer and searching by other determinants may be difficult for the user. Although this type of database is simple in its structure, expanding the number of fields usually entails reprogramming. Additionally, adding new records is time consuming, particularly when there are numerous fields (Figure 4.1).

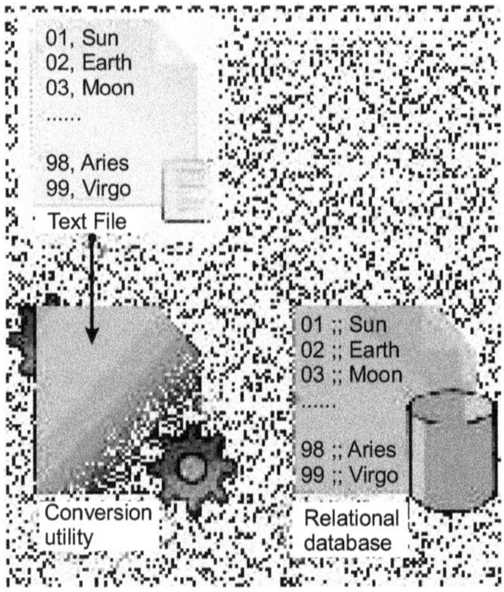

Figure 4.1 A flat file database model

Hierarchical files Hierarchical files store data in more than one type of record. This method is usually described as a parent-to- child or one-to-many relationship. One field is a key to all records but data in one record does not have to be repeated in another. This system will allow records with similar attributes to be associated together. The records are linked to each other by a key field in a hierarchy of files. Each record except for the master record has a higher-level record file linked by a key field pointer. In other words, one record may lead to another and so on in a relatively descending pattern. An advantage is that when the relationship is clearly defined and queries follow a standard routine, a very efficient data structure results. One of the disadvantages is that one must access the master record, with the key field determinant, in order to link downwards to other records (Figure 4.2).

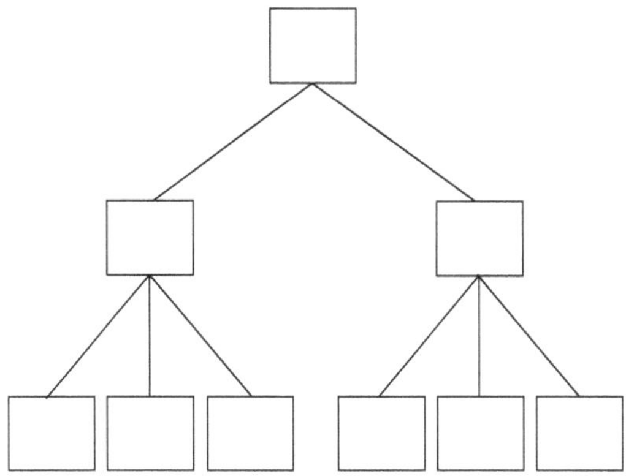

Figure 4.2 A hierarchical file database model

Relational files Relational files connect different files or tables without using internal pointers or keys. Instead, a common link of data is used to join or associate records. The link is not hierarchical. Matrices of tables are used to store the information. As long as the tables have a common link, they may be combined by the user to form new queries and data output. This is the most flexible system and is particularly suited to SQL (Structure Query Language). Because of its flexibility, this system is the most popular database model (Figure 4.3).

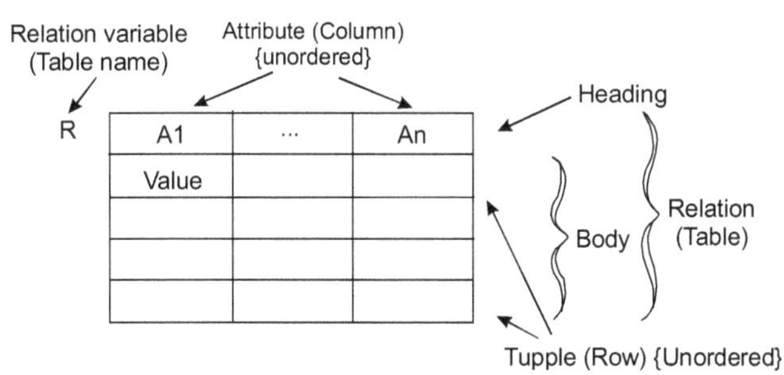

Figure 4.3 A relational file database model

Object-oriented databases Object-oriented database has the advantage of organizing information in ways that we often find it easy to use the database. It has an intuitive feel because it employs those categories which users use naturally in day-to-day life. For this reason, object-oriented databases are gaining increased attention (Figure 4.4).

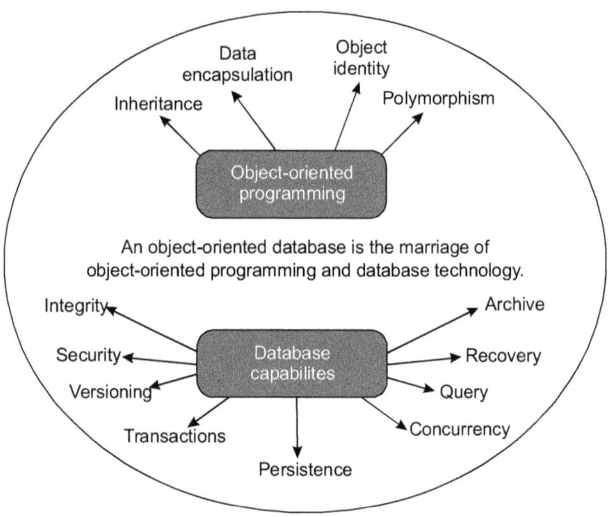

Figure 4.4 An object-oriented database model

IMPORTANCE OF BIOLOGICAL DATABASES

In the last 10 years, both the size of sequence databases and the power of computers has seen exponential growth. Internet has radically affected the way data is provided, handled and analysed. This powerful combination of data and tools, which allow easy access and analysis has changed, and will continue to change our approach to the design and practice of biological research.

One of the major challenges for the academicians and researchers working in the field of bioinformatics is to manage the growing biological resources on the Internet. It is important for the bench scientists working in biology to have easy and efficient ways of wading through the data and finding what is important to his or her research.

Hence biological data is becoming increasingly important to the modern research community. The task of managing this data is important yet the task of presenting this data to the researchers is of even greater importance. Within the last two decades, new technologies have enabled the automated generation of data on a large scale that was never anticipated. The best example of this is the automated DNA sequencing developed by Maxam, Gilbert and Sanger.

MOLECULAR BIOLOGY DATABASES

Recent years have seen an explosive growth in biological data, which is often not published anymore in the conventional sense, but deposited in a database. Sequence data from mega-sequencing projects may not even be linked to a conventional publication. This trend and the need for computational analysis of the data, made databases essential tools for biological research. There are ample specialized databases that it is not reasonable to list the Uniform Resource Locators (URLs) of all of them, especially since this category of databases is quite temporary and any list provided here would be outdated in no time.

Bibliographic Databases

Services that abstract the scientific literature began to make their data available in machine-readable form in the early 1960s. One should be aware that none of the abstracting services has a complete coverage. The best known is MEDLINE and now PUBMED is abstracting mainly the medicinal literature.

MEDLINE and PUBMED can be accessed through NCBI-ENTREZ. EMBASE is a commercial product for the medical literature. BIOSS, the inheritor of the old biological abstracts covers many biological fields like zoological taxonomy, etc. CAB international maintains database in the field of agricultural and parasitic diseases. AGRICOLA is for agriculture, what MEDLINE is for medicine.

Taxonomy Databases

Taxonomy databases are rather controversial since the classification done by one taxonomist will be directly questioned by the other. Various efforts are going on to create a taxonomy resource, e.g. International Organization for Plant Information, Integrated Taxonomic Information System, etc. The most used taxonomy databases are maintained by the NCBI. This hierarchical taxonomy is used by the nucleotide sequence database, SWISS-PROT and TrEMBL, and is curated by an informal group of experts.

Nucleotide Sequence Databases

The International Nucleotide Sequence Database Collaboration (often though inaccurately referred to as GenBank) is a joint production of the nucleotide sequence database by the DDBJ (DNA Data Bank of Japan), EBI (European Bioinformatics Institute) and NCBI (National Centre for Biotechnology Information). In Europe the vast majority of the nucleotide sequence data produced is collected, organized and distributed by EMBL. The nucleotide sequence database is located at the European Bioinformatics Institute, an outstation of the European Molecular Biology Laboratory, (EMBL) in Heidelberg, Germany. The nucleotide sequence databases are data repositories, accepting nucleic acid data from the community and making it freely available. These data are heterogeneous; they vary with respect to the source of the material (e.g. yeast, viruses and cDNA), the intended quality (e.g. finished versus single pass sequence), the extent of sequence annotation and the intended completeness of the sequence relative to its biological target, (e.g. complete versus partial coverage of a gene or a genome).

EMBL, NCBI and DDBJ automatically update each of their entries after every 24 hours with the new sequence they have collected. The result is that they contain exactly the same information, except for sequences that have been added in the last 24 hours.

GenBank is the NIH genetic sequence database, an annotated collection of all publicly available DNA sequences. There are approximately 85,759,586,764 bases in 82,853,685 sequence records in the traditional GenBank divisions and 108,635,736,141 bases in 27,439,206 sequence records in the WGS division as of February 2008.

Established in 1980, the EMBL database was coupled top the publication of sequences in the scientific literature. Electronic submission via the **www** is now usual practice. Today, the majority of data submitted by direct transfer of data comes from the major sequencing centres, such as the Sanger Centre. The EMBL database has nearly tripled in size within a few months and on September 1, 2000 contained more than 9.6 gigabytes of data in 8.3 million records.

DDBJ was started in 1986 at the National Institute of Genetics (NIG) with the endorsement of Ministry of Education, Science, Sports and Culture. From the beginning, DDBJ has been functioning as one of the international DNA databases, including EMBL in Europe and NCBI (GenBank) in the USA as the two major centres for Biotechnology information. Consequently this has been collaborating with the two databanks through exchanging data and information on Internet as the international databanks collaboration.

Each entry in a database will have a unique identifier that is a string of letters and/or numbers that only a specific record has. This unique identifier, which is known as the accession number can be quoted in the scientific literature, as it will never change. Since the accession number must always remain the same, another code is used to indicate the different versions due to sequence corrections. One should therefore always take care to quote both the unique identifier and the version number when referring to records in a nucleotide sequence database. The accession number line in a nucleotide sequence record lists the accession numbers associated with the entry. The accession numbers consist of one letter followed by the five digits (X12345) or (more recently) two letters followed by six digits (XY 123456).

The SV (Sequence Version) line contains the nucleotide sequence identifier, which allows one to recognize the sequence version of this record. An example of a sequence version line is SV AJ000012.1. The nucleotide sequence identifier is of the form of accession version, (e.g. AJ000012.1).

Sequence Cluster Databases

Although the nucleotide sequence data is checked for integrity and obvious errors by the data library staff, the quality of the data is the responsibility of the submitter. As a consequence, there are many errors in the database; many sequence entries are mislabelled, contaminated incompletely or erroneously annotated or contain sequencing errors. In addition to this the database being redundant, in the sense that the same sequence from the same organism may be included many times, simply reflecting the redundancy of the original scientific reports. Sequence cluster databases such as Unigene and STACK (Sequence Tag Alignment and Consensus Knowledge base) address the redundancy problem by coalescing sequences that are sufficiently similar, that one may reasonably infer that they are derived from the same gene.

Several specialized sequence databases are also available. Some of these deal with particular classes of sequences, e.g. the Ribosomal Database Project (RDP), the HIV sequence database and IMGT, the immunogenetics database. Others are focusing on particular features, such as TRANSFAC for transcription factor and transcription factor binding sites and EPD (Eukaryotic Protein Database) for promoters and REBASE for restriction enzymes and restriction enzyme sites. GOBASE is a specialized database of organelle genomes. A database for mitochondrial genomes is Mitbase from EBI.

Genome/Genetic Databases

For organisms of major interest to geneticists, there is a long history of conventionally published catalogues of genes or mutations. In the past few years most of these have been made available in an electronic form and a variety of new databases have been developed. These databases vary greatly in form and content, varying in the classes of data captured and how these data are stored.

There are several databases for *Escherichia coli*. The Coli Genetic Stock Center (CGSE) maintains a database of *E. coli* genetic information including genotypes and reference information for the strains in the CGSC collection, gene names, properties and linkage maps, gene product information and information on specific mutations. The *E. coli* database collection (ECDC) is another example. The Encyclopedia of *E. coli* Genes and Metabolism is a database of *E. coli* genes and metabolic pathways.

The MIPS yeast database is important for information on the yeast genome and its products. The *Saccharomyces* Genome Database (SGD) is another major yeast database.

AceDB is the database for genetic and molecular data concerning *Caenorhabditis elegans*. The database management system written for AceDB by Durbin and J. Thierry-Mieg has proved very popular and has been used in many other species-specific databases. However, it is different from the *C. elegans* database.

Two of the best curated genetic databases are Flybase, the database for *Drosophila melanogaster* and the Mouse Genome Database (MGD). ZFIN, a database for another important model organism, the zebra fish, *Brachydanio rerio*, has been implemented recently. Two major databases for human genes and genomics are in existence. McKusicks Mendelian Inheritance in Man (MIM) which is a catalogue of human genes and genetic disorders is available in an online form (OMIM) from the NCBI. The Genome Database (GDB) is the major human database including both molecular and mapping data. Both OMIM and GDB include information on genetic variation in humans but there is also the human mutation server at the EBI; and to the SRS interface to many human mutation databases.

Protein Sequence Databases

The protein sequence databases are the most comprehensive sources of information on proteins. It is necessary to distinguish between the universal database covering proteins from all species and specialized data collections storing information about specific families or group of proteins or about the protein of a specific organism. Two categories of universal protein sequence database can be seen: simple archives of sequence data and annotated database where additional information has been added to the sequence record. The Protein Information Resource (PIR), the oldest protein sequence database and SWISS-PROT, the annotated universal protein sequence database and TrEMBL, the supplement of SWISS-PROT, can be classified as computer-annotated sequence repositories.

SWISS-PROT is an annotated protein sequence database established in 1986 and maintained collaborately since 1987 by the department of Medical Biochemistry of the University of Geneva and the EMBL data entry library (Now the EMBL can also be called as EBI).

The Protein Information Resource (PIR) in collaboration with Munich Information Center for Protein Sequences (MIPS) and Japan Information Database (JIPID), produces the PIR International Protein Sequence Database (PIR-PSD); a comprehensive, non-redundant, expertly annotated, fully classified and extensively cross-referenced protein sequence database in the public domain. The primary objective for its continuing development and enhancement is to achieve the properties of comprehensiveness, timeliness, non-redundancy, quality annotation and full classification.

Structure Database

The Protein Data Bank (PDB) shows that the number of known protein structures is increasing very rapidly. The Nucleic Acid Database (NDB) is the database of structures of small molecules of interest to biologists concerned with protein–ligand interactions. Another example is the MMDB (Molecular Modelling Database) which stores protein structures derived by molecular modelling.

PDB is the single international repository for public data on the 3-dimensional structures of biological macromolecules. The contents are primarily experimental data derived from X-ray crystallography and NMR experiments. The primary goals of the resource are to enable us to locate a structure of interest, to perform simple analysis on one or more structures or to act as a portal of additional information available on a structure notably the Cartesian atomic coordinates for further analysis.The database is constantly updated as new structures are deposited by the international scientific community.

It was established at the Brookhaven National Laboratories (BNL) in 1971 as an archive for biological macromolecular crystal structures. The archive is updated each year with a handful of structures. In 1980, the number of deposited structures began to increase dramatically. This was due to the improved technology for all aspects of the crystallographic processes.

The Structural Classification of Proteins (SCOP) database aims to provide a detailed and comprehensive description of the structural and evolutionary relationships between all proteins whose structure is known, including all entries in the PDB. It is available as a set of tightly linked hypertext documents which makes the large database comprehensive and accessible. Proteins are classified to reflect both structural and evolutionary relatedness. Many levels exist in the hierarchy but the principal levels are family, superfamily and fold.

CATH is a novel hierarchical classification of protein domain structures, which clusters proteins at four major levels, Class(C), Architecture (A), Topology (T) and Homologous superfamily (H). Class, which is derived from secondary structure content is assigned for more than 90% of protein structures automatically. Architecture, which describes the gross orientation of secondary structures, independent of connectivities, is currently assigned manually. The topology level clusters show their topological connections and the number of secondary structures. The homologous superfamilies clusters proteins with highly similar structures and function. The assignments of structures of topology families and homologous superfamilies are made by sequence and structure comparisons.

Protein Cluster Databases

The concept of superfamilies of proteins was elaborated by Dayhoff[2] and Zuckerkandl[3] following the early observation that many proteins share sequence similarity, suggesting their common evolutionary origin. A very successful attempt to cluster protein sequences is the Pfam database of protein domain families. It contains mutually curated automatic components.

Pfam is a database of multiple alignments of protein domains or conserved protein regions. The alignments represent some evolutionary conserved structures, which has implications for the protein's function. Profile Hidden Markov Models (Profile-HMM) built from the Pfam alignments can be very useful for recognizing that a new protein belongs to an existing protein family, even if the homology is weak. Pfam deals sensibly with multidomain proteins. It is composed of two parts: the first part, Pfam-A contains curated families each with an associated Profile Hidden Markov Model that can be used for alignment and database searching. The second part is Pfam-B, in which sequence segments that are not included in Pfam-A are clustered automatically, allowing Pfam to be comprehensive.

Protein Families and Domains Databases

PROSITE is a database used to determine the function of uncharacterized proteins translated from genomic or cDNA sequences. It consists of a database of biologically significant sites and patterns, formulated in such a way that with appropriate computational tools, it can rapidly and reliably identify as to which family of proteins the new sequence belongs to. To make use of the PROSITE patterns and profiles, one can make use of either of the two software tools, Scan prosite or Profile scan.

Integrated Databases

INTERPRO is an integrated documentation resource for protein families. Domains and sites, developed initially as a means of rationalizing the complementary efforts of PROSITE, PRINTS, Pfam and PRODOM. Many other links too, point to the relevant member database, allowing the user to see at a glance whether a particular family or domain has associated patterns, profiles, fingerprints, etc. Merged and individual entries are assigned unique accession numbers. Each InterPro entry list contains all the matches against SWISS-PROT and TrEMBL.

InterPro aims to reduce duplication of efforts in the labour-intensive, rate-limiting process of annotation and will facilitate communication between the disparate resources. [By uniting these databases, their individual strengths are capitalized, producing a single entity that is by far greater than the sum of its parts.] InterPro will streamline the analysis of newly determined sequences for the individual user and will make a significant contribution in the demanding task of automatic annotation of predicted proteins from genome sequencing projects as it evolves.

INTELLIGENT DATA ORGANIZATION

Comparison is the most commonly used methodology in computational molecular biology. Many biological objects come in families that share structural or functional features. In the early days of bioinformatics, protein sequence and structure comparison were made on a case-by-case basis. More recently many systematic efforts to organize protein-related information in the form of data collections with value additions have been undertaken.

DATA QUALITY AND CONTROL BY DATABANKS

Databanks at two extremes function as passive data repositories (archives) or as active reference compendia, issuing modification of data and information content. Data in biological databanks contains facts, e.g. representations of biological macromolecules as strings or coordinates and associated information, which might be fuzzy, incomplete of subject to individual interpretation or conflicting nomenclature. Data quality has several elements—correctness, completeness, timeliness of capture—applied both to the newly measured properties and the annotation.

Quality control should not be restricted to the semantic checking of individual entries but should also include relationships to the other parts of the databases. The growth in rates of data generation has implication for data quality. On one hand, most new data entries are related to previously described objects and can inherit some part of their annotations. However, many newly determined data entries have no associated experimentally confirmed facts and their annotations are based on their prediction.

INTEGRATED DATA RETRIEVAL SYSTEM

An early attempt to provide a comprehensive interface to a variety of molecular biology databases was made by

George and Orcutt in their ATLAS system. ATLAS generated data indices by parsing the heterogeneous formats into common fields allowing for cross-database multiterm queries, avoiding the need for reformatting the database sources.

Several highly sophisticated retrieval systems have emerged over the recent years. Entrez is a gateway to the data collections, maintained by the NCBI. It includes nucleic acid and protein sequence data, 3D structures, genomes, taxonomic information and the only freely available general literature database, PUBMED. Perhaps the most powerful and unique feature of the Entrez system is that in addition to the static cross-reference inherent to the underlying database, extraction of related documents is also retrieved from the neighbouring document. For example, it is possible to retrieve bibliographic references related to the article you are currently looking at. This is done through lexicographical analysis of the abstract text and keywords. Sequence neighbours can be extracted by performing a keyword search. Sequence neighbours can be extracted by performing a BLAST similarity search. Entrez also has excellent graphical capabilities and is directly connected with a protein structure viewer and an extensive genome browser.

DATABASE PROGRAMS

A database program has tools to:

- design the structure of your database

- create the data entry forms so that one can key in information into the databases
- validate the data entered and check for inconsistencies
- sort and manipulate the data in the database
- query the database (that is, ask questions about the data)

It produces flexible reports, both on screen and on paper that makes it easy to comprehend the information stored in the database. Most of the more advanced database programs have built-in programming or macro languages, which let us automate many of its functions.

To create and maintain a database, one needs a computer database program called a database management system (DBMS). Just as databases range from simple, single-line table lists to complex multiple systems, database systems too range in complexity. Some database components of Microsoft are designed purely to manage single-file databases. With such a product one cannot build a multi-table database. Such programs are sometimes called flat file databases or list managers.

Other database programs, called relational database programs or RDBMS, are designed to handle multi-file databases. FilemakerPro is a relational database, which is easy to use and fairly inexpensive.

The most popular relational databases are offered from three big software companies. Lotus, Corel and Microsoft each produces a full-featured relational database application available both as a standalone program and as a part of its integrated suite; Lotus has Approach, Coral has Paradox and Microsoft has Access.

TEXT-BASED SEARCHING

Some of the data retrieval systems of particular relevance to molecular biologists are Sequence Retrieval System (SRS), ENTREZ and DBGET. These three systems allow searching in a multitude of molecular biology databases and provide links to relevant information for entries that match the search criteria. The three systems differ in the databases they search and the links they have to other information.

SRS

This sequence retrieval system is a homogeneous interface for over 80 biological databases that have been developed at EBI. It includes databases of sequences, metabolic pathways (BLAST, SSEARCH, and FASTA), transcription factors, and application results like protein 3D structures, genomes, mappings, mutations and locus-specific mutations. The webpage listing of all the databases contains a link to a description page about each database including the data on which it was last updated. One can select one or more of the databases to search before entering one's query. After getting the result, one can choose an alignment algorithm (like ClustalW, Phylip), enter parameters and run it. The SRS is highly recommended for use.

ENTREZ

This is a molecular biology database and retrieval system, developed by NCBI. It is the entry point for exploring distinct but integrated databases. Of the three texts-based database systems, Entrez is the easiest to use but also offers limited information to search. Entrez is a combined bibliographic, protein sequence and nucleotide sequence database maintained by NCBI. A major advantage of Entrez is interconnections between the various databases; one can move quickly from a sequence to its reference to another sequence. Another central concept in Entrez is neighbouring, the grouping together of sequences and references by computed similarity scores. Entrez has sprouted several variants. First, the Entrez database can be accessed by either a CD-ROM or over the Internet. Second, Entrez can be used with either a custom graphical interface client, using a World Wide Web browser, with a command line browser or via NCBI's toolkit written in C language. Many biosequence databases are available as hypertext on the World Wide Web or as flat files from Gophers.

DBGET

This is an integrated database retrieval system developed by the Institute for Chemical Research, Kyoto University and the Human Genome Center of the University of Tokyo, which provides access to 20 databases one at a time having more options. The DBGET is less recommended than the other two.

SEARCH CONCEPTS

Boolean Search

This is an advanced query technique used when searching with two or more terms. Terms are combined using the Boolean operators AND, OR and NOT. The default Boolean operator is AND.

Broadening the Search

If the results of a search produce no useful entries, the terms that have been entered are either changed or removed.

Narrowing the Search

If the results of a search produce too many entries, the terms that have been entered are changed or added.

Proximity Search

To search with multiword terms or phrases, quotes are placed around the terms.

Wild Cards

The character can be appended to a search term to make a search less specific. For example, to look for all authors whose last name begins with Khan, the word Khan is used.

DATABASE SEARCHING—BASIC CONSIDERATIONS

Most sequence databases have a significant number of cross-references to other databases. A protein sequence, for example, will have one or more references to the DNA sequence coding for the protein and possibly also points to databases describing protein motifs (such as the PROSITE database) or organism-specific databases. Recently, the interest of researchers focuses on genome projects.

To make the best use of the widely available databases, one first needs to find out which databases are storing the information we are looking for in the most comprehensive fashion. If one searches only for a given accession number, one will be able to search all the sequence databases simultaneously. However, searching a genetic locus of a disease or a protein motif for a specific protein function will succeed more efficiently if we use one of the databases specifically made for this purpose. In the two examples mentioned, the databases of choice are OMIM and PROSITE, respectively. When we have found one description of a sequence, our search is not finished. Once one encounters hits in one database, one should use this information to search other databases as well.

The access to databases is no longer necessarily performed on the same computer where one usually does sequence analysis. Some programs operate via networks WWW browser. It is, however, important to note that the retrieved sequences will be in specific formats. The data will be ordered in a way that the software one wants to use for further analysis can or cannot interpret them correctly. Therefore, we must determine the formats of the entries that get via computer networks and apply appropriate procedures for reformatting, if the data shall be used in the GCG program package.

Security Notice

When we use wide area computer networks, we will most probably access databases and computers, which are not under local control. Information quality, therefore, might not apply in the usual way. This consideration is particularly important for environments beyond firewalls (commercial software). The two main ways of searching are: Text-based search, querying the annotation and sequence-based search.

SOME WEBSITES FOR DATABASE SEARCHES

BLOCKS-http://www.blocks.mere.org/blocksmkr/makeblocks.html

Profiles-http://www.psc.edu/general/software/packages/profiles/profile ss.html

PROBE-ncbi.nlm.nih.gor/pub/newwald

BCM-http://dot.imgen.bcm.tmc.edu.9331/seq.search/protein.search.html

FASTA-http://jasta.bioch.virginia.edu/FASTA

TIGR-http://www.expasy.ch/prosite

INTERPRO-www.ebi.ac.uk/interpro

PFAM-http://www.Sanger.al.uk/pfam

PSI-BLAST-http://www.ncbi-nm.nib.gar\BLAST

ENTREZ-http://www.ncbi.nlm.nih.gov/Entrez.

Tree of Life-http://Phylogeny. arizona. edu/tree/ phylogency.htm

PDB-http://www.rcsb.org

SCOP-http://scop mre-lmb.cam.ac.uk/SCOP

Cn3d-http://www.ncbi.nlm.nih.gore/structural /CN3D

Rasmol-http:// www.umass.edu/microbiol/rasimol

REVIEW QUESTIONS

1. Describe the various data models with illustrations to explain each of them.
2. Write a detailed account of various nucleic acid databases.
3. Give an account of protein sequence databases.
4. Describe a few protein structure databases quoting examples.
5. List the methods of text-based searching.

REFERENCES

1. Michael, R. McCaleb. (1999). "A conceptual data model of datum systems." National Institute of Standards and Technology. August 1999.

2. Dayhoff, M.O., Barker, W.C., Schwartz, R.M., Orcutt, B.C. and Hunt, L.T. (1976). "Data Base for Protein Sequences." AFIPS National Computer Conference Proceedings. Vol. 45, 261–266 .
3. Zuckerkandl, E. (1975). "The appearance of new structures and functions in proteins during evolution." *J. Mol. Evol.* 7(1): 1–57.

Websites

1. www.w3.org/WhatIs.html
2. www.un.org/databases
3. www.ncbi.nlm.nih.gov/Database/
4. www.webopedia.com/TERM/d/database.html
5. www.nist.gov/srd/
6. www.eh.net/databases/
7. www.nlm.nih.gov/medlineplus/databases.html

5

SEQUENCE ANALYSIS

We didn't know it at the time, but we found everything in life is so similar, that the same genes working in flies are the ones that work in humans.

Eric Wieschaus

OBJECTIVES

1. To enable understanding of the function of the DNA sequences and proteins
2. To perform DNA or protein sequence alignment in order to analyse their relatedness as well as their evolutionary origin

INTRODUCTION

In biology, sequence analysis implies subjecting a DNA or protein sequence to sequence alignment, sequence databases, repeated sequence searches, or other bioinformatics methods on a computer. It is an automated, computer-based examination of characteristic fragments, which encompasses the use of various bioinformatics methods to determine the biological function and/or structure of genes and the proteins they code for. The DNA or protein sequences are first compared with the known sequences in the database.

Comparison is done at the level of constituent amino acid residues/nucleotides, in order to find out conserved variable sequence patterns between the new and the known sequences. Presence of conserved residues especially those relevant to protein function would enable one to guess about the nature of the protein. In order to find out the conserved residues, the residues of one of the sequence are directly mapped on to the residues of the other sequences. The process of mapping is called sequence alignment. It includes five biologically relevant topics.

1. Comparison of sequences in order to find similar sequences (sequence alignment)
2. Identification of gene-structures, reading frames, distribution of introns and exons and regulatory elements
3. Prediction of protein structures

4. Genome mapping
5. Comparison of homologous sequences to construct a molecular phylogenetic tree

SEQUENCE ALIGNMENT REPRESENTATIONS

Sequence alignments are commonly represented both graphically and in text format. The sequences are written in rows arranged so that aligned residues appear in successive columns. In text formats, aligned columns containing identical or similar characters are indicated with symbols. An asterisk or pipe symbol is used to show identity between two columns, a colon is used for conservative substitutions and a period for semi-conservative substitutions.

Many sequence visualization programs also use colour to display information about the properties of the individual sequence elements in DNA and RNA sequences. This is similar to assigning each nucleotide its own colour. In protein alignments, colour is often used to indicate amino acid properties to aid in judging the conservation of a given amino acid substitution. In multiple sequences, the last row in each column is often the consensus sequence determined by the alignment; the consensus sequence is also often represented in graphical format with a sequence logo in which the size of each nucleotide or amino acid letter corresponds to its degree of conservation.[1]

SEQUENCE ALIGNMENT AND ALIGNMENT PROGRAMS

Sequence alignment is a way of arranging the primary sequences of DNA, RNA, or protein to identify regions of similarity that may be a consequence of functional, structural, or evolutionary relationships between the sequences. Aligned sequences of nucleotide or amino acid residues are typically represented as rows within a matrix. Gaps are inserted between the residues so that residues with identical or similar characters are aligned in successive columns.

Alignments can be local or global. In global alignment, whole sequences are considered whereas in local alignments, only parts of sequences are considered.[2] In both cases, the basic goal is to achieve an alignment, which gives rise to maximum number of matches (i.e., high sequence similarity). Such an alignment is known as optimum alignment. Both the type of alignments have advantages as well as disadvantages and are applied based on the type of application. For example, in comparative modelling global sequence alignment is essential whereas local alignment is sufficient to detect functional domains. Computer software like ClustalX and MALIGN do global alignments and heuristic software like BLAST, FASTA do local alignments. Global alignment requires more time computationally than the local alignment. Typically one would use local alignment for database searches for detecting sequentially similar sequences and global alignments for performing in-depth analysis on residue conservation pattern, residue mutation and residue deletion and insertion.

COMPARISON OF TWO SEQUENCES—SCHEMATIC COMPARISON

Principle of Sequence Alignment

Let us assume the following two sequences:

Known sequence: tgatggtcaagtaaactatgaagagtt

Unknown sequence: atggtaatggcacaattgactttcctgaatttctga

If we want to align these, we will try to write the two sequences in a way, which allows a pairwise comparison of the sequences. There are a lot of possible options to do so and the longer the sequences, the more options to align two sequences. In order to find the best alignment, we need to judge the quality of the alignment. To allow computations and comparisons, this judgment shall result in a numerical value, which is called a **score**. The determination of this score relies on a symbol comparison table, where each symbol pairing gets a value assigned, in order to determine the overall score by adding up the comparison value of each observed pair in the alignment. These tables are very important for proteins, but are also used in DNA comparison. A typical, simple scoring table for nucleotides will give a value of 1 to a match (treating U and T as match) and assign a value of 0 to each mismatch:

Match value: 1
Mismatch value: 0

	A	G	C	T	U
A	1	0	0	0	0
G	0	1	0	0	0
C	0	0	1	0	0
T	0	0	0	1	1
U	0	0	0	1	1

Figure 5.1 Scoring table

This matrix is perfectly symmetric and would be sufficient if printed as a half-populated table.

To get started with, we will write the two sequences one below the other. However, we can either align the beginning, the end, or arrange the sequences arbitrarily. The score is determined according to the table shown in Figure 5.1. Figure 5.2 shows that the two sequences are shifted at various positions. Sequence alignments are produced by shifting. Scores are calculated using a match value of 1 and a mismatch value of 0. The mismatch value is usually given as –0.5.

Figure 5.2 Match and mismatch

Two sequences can always be aligned, irrespective of whether they are similar or not. It is possible to score sequence alignments and use the obtained numbers as a measure to discriminate between several alignments. Alignments will possibly yield more than one possible solution with similar score. The scoring table, if written as a best-score listing of the top four alignments, will read as:

Score	Shift	Length
15	–4	29
10	2	27
	–5	29
9	3	26

Figure 5.3 Scoring table

This means that one alignment with shift –4 is calculated to be best, but the alignments with shift 2, –5 and 3 are of a similar score.

This type of scoring will favour long alignments and will produce higher scores for the longer alignments. However, the mismatches are not penalized, which implies that long stretches of different sequences might be in the alignment. The result will be that the score gets better if the alignment gets longer, regardless of the amount of mismatches encountered. In order to discriminate better between similar sequences and those which have accidental similarity on a long range of symbols (such as expected in G/C-rich sequences), we need to change the scoring to penalize mismatches. As an example, we use the scoring to recalculate scores. Figure 5.3 shows the scoring table, if written as best score listing of the top four alignments.

Match value : +1.0

Mismatch value : –0.5

Dot Matrix Analysis

This is a graphical method, primarily used to find regions of global matches between two sequences. This method was first introduced by Gibbs and Mclntyne (1970) (Figure 5.4) and is very simple method. The two sequences to be compared are placed as rows and columns of matrix and the matrix is filled with dots.

There are a number of programs available for dot matrix analysis. Some programs are as follows:

1. DOTTER
2. PALIGN
3. DOTLET

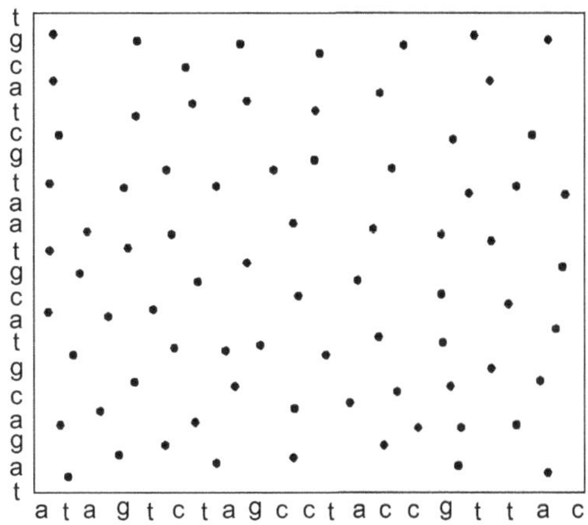

Figure 5.5 Dot-plot

Number of possible dots = (Probability of word) * (length of sequence A) * (length of sequence B)

SCORING SCHEMES AND SUBSTITUTION MATRICES

To work out an optimum alignment computationally which maximizes similarity between sequences, we need to use some scoring methods. For example, we can use scores based on residue-to-residue matching. A score of 1 for match and 9 for mismatch can be assigned. Therefore we could write a scoring matrix as identity matrix. Similarly, we can write 20*20 identity matrix for 20 amino acid residues with all the diagonal elements carrying a score of 1 and all non-diagonal elements carrying a score of 0. Identity matrix can be used for sequence alignments. However, we need to use more robust scoring schemes. Evolutionarily it has been observed that residues often get substituted or mutated without causing any harm to the native structure and function of the protein. The residues involved in such mutations are considered similar. For example, substitutions like isoleucine to valine is hydrophobic and serine to threonine is polar. Researchers have analysed many homologous protein sequences and have noted that the residue pairs are similar. They have also found that residues do show mutational preferences (i.e., only certain pairs most often get interchanged), probably due to evolutionary pressure as well as structural and functional restraint. As a consequence of mutational preferences, some pairs of residues become more similar than the other pairs. For computational purposes, residue-to-residue similarity is expressed in terms of similarity scores. A 20*20 matrix of similarity scores is known as "similarity matrix" (also called as a substitution matrix).

A number of similarity matrices have been derived considering various similarity features of the residues. The most widely used ones are discussed below.

PAM Matrices

The first substitution matrix, which gained very wide usage and acceptance, as devised by Dayhoff and co-workers (1978). These are based on the Point Accepted Mutation (PAM) model of evolution. This model assumes that the evolutionary changes occur as Markov Models, i.e., residue mutations occur independent of the previous mutations. One PAM is a unit of evolutionary divergence in which 1% of the amino acids have been changed; this does not imply that after 100 PAMs, every amino acid will be different. Some positions may change several times, perhaps reverting to the original amino acid, whereas others may not change at all.

In homologous protein residues, substitutions are subjected to restrictions due to necessity of retaining the native three-dimensional structure and function.[3] In fact certain residues remain conserved all along evolution. For example, the catalytic residues D, T and G are aspartic proteases. The mutations that do not cause serious disruption of protein native structure and function are said to be accepted point mutation. Dayhoff and her colleagues calculated frequencies of accepted mutations for PAM1 by analysing closely related protein sequences (i.e., those homologues which can be aligned manually without the aid of a substitution matrix). The similarity scores were calculated as a natural logarithm of ratio of observed substitution frequencies (referred to as target frequencies). These similarity scores are also referred to as log odds ratio. This method of calculation, first adopted by Dayhoff has been used as a basic for calculating all kinds of similarity scores known till today. The data calculated for PAM1 as extrapolated to a distance of 250 PAMs and the resulting matrix was published in 1978. This matrix has been popularly referred to as PAM20.

The data for PAM1 has been extrapolated to other PAM distances to produce a family of matrices (e.g. PAM200) to suit different alignment situations. For example, to derive PAM100 matrix, PAM1, matrix is multiplied by itself 100 times. In general PAM*n* matrix can be obtained by multiplying PAM1 by itself *n* number of times. As the usage of PAM matrices is considered as a rule of thumb, it has been suggested to use lower distance PAM matrices for closely related protein and higher PAM matrices for more diverged protein sequences, e.g. PAM30 for closely related proteins whereas PAM250 for highly diverged proteins.

BLOSUM Matrices

BLOSUM is the abbreviated form of BLOCKS amino acid substitution matrices. These matrices have been constructed in a similar fashion as PAM matrices. However, the data were derived from local alignments for distantly related proteins deposited in the BLOCKS database. Unlike PAM matrices, there is no evolutionary basis in the data used for computing target frequencies. BLOSUM also has several versions. For example, BLOSUM 62 was derived from a set of sequences which are 62 per cent or less similar.[4] It is a good practice to use appropriate versions of BLOSUM while doing sequence comparisons. For example, BLOSUM 30 should be used for comparing highly diverged sequences and BLOSUM 90 for very closely related sequences.

A similarity matrix based on the assumption that genetic code is the only influencing factor for amino acid substitution was devised by Benner and co-workers. Matrix by Gonnet (1994), which is a 400*400 dipeptide substitution matrix, was derived on the probability that amino acid substitutions at a particular site are influenced by neighbouring amino acids and thus the environment of an amino acid plays a role in protein evolution.

Matrices are used for assigning scores for residue-to-residue substitution. Introduction of gap is not viewed as favourable and it is penalized. The score associated with the gap is generally called a gap penalty.

DYNAMIC PROGRAMMING

Dynamic programming is a computational method used for aligning two protein or nucleotide sequences. The method compares every pair of amino acid/nucleotide sequence and generates an alignment. In the alignment, matches, mismatches and gaps are positioned in such a way that the number of matches between the identical or similar residues is maximum. Dynamic programming method used for global alignment is called as **Needleman–Wunch algorithm** and that used for local alignment is called as **Smith–Waterman algorithm**.

Needleman–Wunch Algorithm

Let us consider that we have two sequences X and Y and residues $x_1, x_2, x_3 \ldots$ and $y_1, y_2, y_3 \ldots$ respectively. The idea is to build an optimal alignment using previous solutions for optimal alignments of smaller subsequences. In order to achieve this, matrix M is constructed such that each element of M, i.e., $M(i, j)$ is the score of the best alignment possible between the segments x_1 to x_i and the segments y_1 to y_j.

Each element $M(i, j)$ is set equal to the maximum of the scores obtained by the following expressions:

$$M(i-1, j-1) + s(xi, yj)$$

$$M(i-1j) + d$$

$$M(i,j-1) + d$$

where $s(xi, yj)$ is the similarity score for the residues xi and yj taken from a given substitution matrix and d is the linear gap penalty.

For example, we have two sequences AGCTTA and CCGA. Let us assume that we have a scoring scheme as follows.

Every match has a score of 10, every mismatch has a score of –1 and the gap penalty is –5. By using the above algorithm we can work out a matrix as shown in Figure 5.5.

	A	A	G	C	T	T	A
A	0	–5	–10	–15	–20	–25	–30
C	–5	–1	–8	0	–5	–10	–15
C	–10	–6	–2	4	–0	–6	–11
G	–15	–11	4	–1	3	–2	–7
A	–20	–5	–1	3	–2	2	8

Figure 5.5 Dynamic programming scoring matrix

To derive an optimum alignment using the matrix M, the trace-back approach is adopted. In this approach the alignment is built in reverse, starting from the final element of the matrix with a highest score of 8. This score corresponds to the total alignment score because this is the last element of the matrix. At each step of the trace-back process a move is made from the current element (i, j) to one of the elements $(i-1, j-1)$, $(i-1, j)$ or $(i, j-1)$ from which the value of $M(i, j)$ was derived. At the same time a pair of symbols is added in front of the current alignment, that is, residue pair xi and yj is added if the trace-back leads to the element $(i-1, j)$ and residue yj is added if the element is $(j, i-1)$.

Smith–Waterman Algorithm

Most often it is sufficient to do alignments of subsequences rather than global alignments from one end to the other end of the sequences. For example, when we compare highly diverged sequences it is always better to do local sequence, alignments, as there can be conserved domains in them. This is because residue matches are stronger in conserved domains than the other parts. In order to identify the subsequences which give rise to maximum match locally, Smith and Waterman introduced two modifications into the dynamic programming as discussed below.

The first modification is in the way of matrix updation carried out. Unlike Needleman–Wunch algorithm, the element $M(i, j)$ is set equal to either the maximum of the scores calculated using the three expressions mentioned above or to zero. The value zero is selected when all the expressions give rise to scores less than zero. The importance of choosing zero is to stop the best alignment at some point when it gives rise to a negative score at that point and start a new alignment.

The second modification is in the trace back procedure followed. Unlike the Needleman–Wunch algorithm, where the trace-back begins from the last element of the matrix, here the trace-back starts from anywhere in the matrix where the element $M(i, j)$ has the highest score.

So far we have assumed the linear gap penalty when deriving the elements of the dynamic programming matrix. The other alternative is to use affine gap score penalty, which is more robust than linear gap score scheme. In the affine gap score penalty, we need to calculate three matrices M, ix and iy instead of one. Here ix and iy matrices explicitly correspond to scores calculated when a gap is introduced for a residue in the sequences x and y respectively. The three matrices are updated as follows:

$M(I-1, j-1) + s(xi, yj)$

$Lx(I-1, j-1) + s(xi, yi)$

$Ly(I-1, j-1) + s(xi, yi)$

$Lx(I, j)$ is set equal to the maximum of the following:

$M(I-1, j) - d$

$Lx(I-1, j) - e$

$Ly(I, j)$ is set equal to the maximum of the following:

$Ly(I, j-1) - e;$

$M(I, j)$ is set equal to the maximum of the following:

$M(I, j\text{–}1) - d$

It should be noted that the trace-back process to deduce optimum alignment involves three matrices.

Heuristic Methods—BLAST and FASTA

The alignment algorithms discussed above are guaranteed to find the optimal alignment according to the scoring scheme adopted for calculations. In particular, the scheme involving affine gap score is generally regarded as providing the most sensitive sequence matches. However, they are not the fastest sequence alignment procedures. For applications like homology searches, using large database speed becomes a prime issue. The dynamic algorithm has time complexity of the order of 09n*m) where n and m are the lengths of the sequences considered for comparisons (i.e., n and m are the number of residues/ nucleotides) to made search in a database containing 100 million residues we need to evaluate about 10^{11} matrix elements.

This may take several hours of computer time. If we want to make homology searches for many sequences, it becomes a very tiring job as far as time is concerned.

In order to reduce the time consumed for comparison, a few heuristic methods have been developed. The word heuristic means "an alternative procedure" mostly based on probably an educated guess. In heuristic methods, there is always trade-off between sensitivity and speed. The most popular methods are BLAST and FASTA.

BLAST BLAST is the abbreviated form of Basic Local Alignment Search Tool. This was proposed by Altschul *et al.* in 1990. BLAST provides software tools for finding high-scoring local alignments between two sequences. The basic assumption adopted by BLAST is that within the segments in a high-scoring local alignment, there exist short stretches of matching segments, i.e., very high scoring segments. Therefore first seed alignments are made between the short matching segments. The lengths of the seed segments are three residues in proteins and 11 in nucleic acids.

BLAST is a popular program for searching biosequences against databases. BLAST was developed and is maintained by a group at NCBI.

Once the seed alignments are made, the ungapped alignments are extended on both the directions up to a certain length where the alignment score stops increasing further. In other words BLAST tries to get high-scoring local alignments from the seed alignments.

One of the attractive features of BLAST is that it tries to report as many high-scoring local alignments as possible for a pair of given sequences. Hence it becomes easy for a researcher to analyse the hits. The high-scoring pairs constituting the query and the subject (i.e., the sequence from the database) are generally called as HSPs (High-Scoring Pairs).

The most widely used implementation of BLAST finds ungapped alignments only. However, gapped version of BLAST is also available. There are several variants of

BLAST each distinguished by the type of sequences (DNA and protein).

BLAST comes in 4 flavours:

BLASTP Searches a protein sequence against a protein database.

BLASTN Searches a nucleotide sequence against a nucleotide database.

TBLASTN Searches a protein sequence against a nucleotide database, by translating each database nucleotide sequence in all 6 reading frames.

BLASTX Searches a nucleotide sequence against a protein database, by first translating the query nucleotide sequence in all 6 reading frames.

Analysis of BLAST reports can be aided by a number of tools such as the filters XNU, SEG, and XBLAST.

A complete set of BLAST programs along with the substitution matrices and non-redundant protein sequence databases and DNA sequence databases are freely available from the FTP file ftp.ncbi.nlm.nih.gov.blast. While doing database searches using BLAST, one of the important things done is to mask the so-called low-complexity regions (LCR). These are the regions of biased composition. These regions can be homopolymeric runs or short-period repeats or subtle cases where one or more residues are over-represented. LCR pose problems while doing database searches, as they do not fit into the model of residue-to-residue sequence conservation. Further more methods for assessing the statistical significance of alignments are based on a certain notion of randomness which LCRs do not obey. Consequently many false positives may be observed in the output of a database search.

The Salient Features of BLAST

1. ***Local alignments*** BLAST tries to find patches of regional similarity, rather than trying to find the best alignment between the entire query and an entire database sequence.
2. ***Ungapped alignments*** Alignments generated with BLAST do not contain gaps. BLAST's speed and statistical model depend on this but in theory it reduces sensitivity. However, BLAST will report multiple local alignments between the query and a database sequence.
3. ***Explicit statistical theory*** BLAST is based on an explicit statistical theory developed by Samuel Karlin and Steven Altschul (1990). The original theory was later extended to cover multiple weak matches between query and database entries (1993).

Both phases of the alignment process (scanning and extension) use a substitution matrix to score matches. This is in contrast to FASTA, which uses a substitution matrix only for the extension phase. Substitution matrices greatly improve sensitivity.

PSI–BLAST A new version of BLAST called Position-Specific Iterated BLAST has been designed to build profiles iteratively and then to do BLAST search. PSI-BLAST involves a series of repeated steps or iterations. First a database search of protein sequence database is performed using query sequence. Second, the results of the search are checked to include only the convincing hits. In other words only the high-scoring sequence matches. These sequences are aligned and a weight matrix is produced from the alignment. The database is again searched with this scoring matrix. Again the hits are aligned and the weight matrix is updated. This type of database search and updating of weight matrix are repeatedly carried out until no new hits are found (converged).

PSI-BLAST is highly sensitive as well as specific due to the use of family-specific profile which gets updated iteratively and that is used as the scoring matrix for the next database search. An analysis of the performance of PSI-BLAST has demonstrated that BLAST can detect weak similarities that exist between distant homologues.

BLAST3 It is also worth mentioning the program BLAST3. This searches a protein against a protein databank using the BLAST algorithm (with the sensitivity set high) and then makes threefold alignments between the query sequence and each possible pair of databank sequences that have been found. Only the statistically significant threefold alignments which are made from three non-significant pairwise alignments are retained. BLAST3 is useful in finding proteins that share a region of only weak similarity. Occasionally, it can show that a query sequence makes the bridge between two databank sequences whose relationship had not yet been suspected.

FASTA The first widely used database searches were FASTA and this was developed by Lippman and Pearson in 1985. A comparative assessment of the performances of BLAST and FASTA has shown that FASTA is more sensitive than BLAST.

In the step (1), FASTA uses a lookup table to locate all identically matching words of length tuples between the two sequences. For proteins ktup is typically 1 or 2 and for DNA it may be 4 or 6. It then looks for diagonals with mutually supporting word matches. This is a very fast operation, which for example can be done by sorting the matches on the difference of indices (i–j). The best diagonals are pursued further in step (2), which is analogous to the hit extension step of BLAST algorithm, extending the exact word matches to find maximal scoring ungapped regions. Step (3) then checks to see if any of these ungapped regions can be joined by a gapped region allowing gap costs. In the final step, the highest scoring candidate matches in a database search are realigned using the full dynamic programming matrix forming a band around the candidate heuristic match.

Because the last step of FASTA uses dynamic programming, the scores it produces can be handled exactly like those from full dynamic programming. One can choose between the speed and sensitivity by choosing ktup values. ktup sensitivity increases with high values, however, the search becomes slow.

THE STATISTICS OF SEQUENCE-SIMILARITY SCORES

To assess whether a given alignment constitutes evidence for homology, it helps to know how strong an alignment can be expected from change alone. In this context, chance can mean the comparison of

i. real but non-homologous sequences;
ii. real sequences that are shuffled to preserve compositional properties; or
iii. sequences that are generated randomly based upon a DNA or protein sequence model.

Analytic statistical results invariably use the last of these definitions of change, while empirical results based on simulation and curve fitting may use any of the definitions.

The Statistics of Global Sequence Comparison

Unfortunately, under even the simplest random models and scoring systems, very little is known about the random distribution of optimal global alignment scores. Monte Carlo experiments can provide rough distributional results for some specific scoring systems and sequence compositions but these cannot be generalized easily. Therefore, one of the few methods available for assessing the statistical significance of a particular global alignment is to generate many random sequence pairs of appropriate length and composition, and calculate the optimal alignment score for each. While it is then possible to express the score of interest in terms of standard deviations from the mean, it is a mistake to assume that the relevant distribution is normal and convert this Z-value into a P-value; the tail behaviour of global alignment score is unknown. 100 random alignments have a score inferior to the alignment of interest. The P-value in question is likely less than 0.01. One further pitfall to avoid is exaggerating the significance of a result found among multiple tests.

When many alignments have been generated, e.g. in a database search, the significance of the best must be discounted accordingly. An alignment with P-value 0.0001 in the context of a single trial may be assigned a P-value of only 0.1, if it was selected as the best among 1000 independent trials.

The Statistics of Local Sequence Comparison

Fortunately statistics for the scores of local alignments, unlike those of global alignments, are well understood. This is particularly true for local alignments lacking gaps, which we will consider first. Such alignments were precisely those sought by the original BLAST database search programs. A local alignment without gaps simply consists of a pair of equal-length segments, one from each of the two sequences being compared. A modification of the Smith–Waterman or Sellers algorithm will find all segment pairs. Those scores cannot be improved by extension or trimming. These are called High-scoring Segment Pairs (HSPs).

To analyse how high a score is likely to arise by chance, a model of random sequences is needed. For proteins, the simplest model chooses the amino acid residues in a sequence independently, with specific background probabilities for the various residues. Additionally, the expected score for aligning a random pair of amino acids is required to be negative. Long alignments would tend to have high score independently of whether the segments aligned were related and the statistical theory would break down.

Just as the sum of a large number of independent identically distributed (i.i.d.) random variables tends to a normal distribution, the maximum or a large number of i.i.d. random variables tend to an extreme value distribution. In studying optimal local sequence alignments, we are essentially dealing with the latter case. In the limit of sufficiently large sequence lengths m and n, the statistics of HSP scores are characterized by two parameters, K and E . Most simply, the expected number of HSPs with a score of at least S is given by the formula:

$$E = Kmn\,e^{-\lambda s} \tag{5.1}$$

we call this the E-value for the score S.

This formula makes eminently intuitive sense. Doubling the length of either sequence should double the number of HSPs attaining a given score. Also for an HSP to attain a score of $2x$, it must attain the score x twice in a row, so one expects E to decrease exponentially with the score.

Bit scores Raw scores have little meaning without detailed knowledge of the scoring system used or more simply its statistical parameter K. Unless the scoring system is understood, citing a raw score alone is like citing a distance without specifying feet, metres or light years. By normalizing a raw score using the formula.

$$S' = \frac{\lambda S - \ln K}{\ln 2} \tag{5.2}$$

one attains a bit score S', which has a standard set of units. The E-value corresponding to a given bit score is simply

$$E = mn2^{-S'} \tag{5.3}$$

Bit scores assume the statistical essence of the scoring system employed, so that to calculate significance one needs to know in addition only the size of the search space.

P-values The number of random HSPs with score of S is described by a Poisson distribution. This means that the probability of finding exactly a HSP with score S is given by

$$e^{-E}\frac{E^a}{a!} \tag{5.4}$$

where E is the E-value of S given by equation (5.1) above. Specifically the chance of finding zero HSPs with score S is e^{-E}; so the probability of finding at least one such HSP is

$$P = 1 - e^{-E} \tag{5.5}$$

This is the P-value associated with the score S. For example, if one expects to find three HSPs with a score a S, the probability of finding at least one is 0.95. The BLAST programs report E-value rather than P-values because it is easier to understand the difference.

For example, E-value of 5 and 0 than P-values of 0.993 and 0.99995. However, when $E<0.01$, P-values and E-values are nearly identical.

The Statistics of Gapped Alignments

The statistics developed above has a solid theoretical foundation only for local alignments that are not permitted to have gaps. However, many computational experiments and some analytic results strongly suggest that the same theory applies as well to gapped alignments.

For ungapped alignments, the statistical parameters can be calculated, using analytic formulae, from the substitution scores and the background residue frequencies of the sequences being compared. For gapped alignments, these parameters must be estimated from a large-scale comparison of random sequences.

Gap Scores

Our theoretical development concerning the optimality of matrices constructed using equation (5.6) unfortunately is invalid as soon as gaps and associated gap scores are introduced, and no more general theory is available to take its place. However, if the gap scores employed are sufficiently large, one can expect that the optimal substitution scores for a given application will not change substantially. In practice, the same substitution scores have been applied fruitfully to local alignments both with and without gaps. Appropriate gap scores have been selected over the years by trial and error and most alignment programs will have a default set of gap scores to go with a default set of substitution scores. If the user wishes to employ a different set of substitution scores, there is no guarantee that the same gap scores will remain appropriate. No clear theoretical guidance can be given, but affine gap scores, with a large penalty for opening a gap and a much smaller one for extending it, have generally proved among the most effective.

MULTIPLE SEQUENCE ALIGNMENT (MSA)

The purpose of multiple sequence alignment is to bring the large number of similar features in the same column of the alignment, optimally. If the sequences in the MSA (Multiple Sequence Alignment) align well, they are likely to be derived from a common ancestor sequence. Presence of several similar domains in several sequences suggests a biochemical function which may become the basis of further experimental investigation.

Based on similarities obtained by MSA, similar proteins have been organized into databases of protein families. Multiple sequence alignment of a set of sequences can provide information as to the most alike regions in the set. In proteins, such regions may represent conserved functional or structural domains.[4]

A good alignment is that which includes a series of columns where majority of sequences have the same amino acid or an amino acid that is a conservative substitution for that amino acid. In the same column very few examples of other substitutions or gaps may be present. These columns should be present throughput the alignment, often clustered into domains. In the case of nucleic acids, multiple sequence alignment can reveal structural and functional relationships. When promoter regions of a set of similar sequences are aligned well, they may represent consensus binding sites for regulatory proteins.[5]

Classes of Multiple Sequence Alignment

(a) *Global alignment* Global sequence alignment is the entire length sequence alignment. It is an extension of dynamic programming algorithm.

(b) *Sequence block* Sequence block is a common pattern of alignment in a group of sequence where matches and mismatches, but not gaps, are included. Sequence blocks can be found by pattern matching or pattern-finding algorithms.

(c) *Profile* A profile is a type of scoring matrix, which is produced from an alignment of common patterns in protein sequences that include matches, mismatches, insertions and deletions. Multiple sequence alignments can be scored by using SP (sum of pairs) model for scoring.

With SP model we can score the MSA by adding all the possible combination of pairs of scores of amino acids in a column of an MSA. This model assumes that any sequence could be an ancestor of the other sequence. When the number of mismatched residue pairs increases, scores in the MSA column decrease rapidly.

SP method is an optimization method where maximization of number of matched pairs score is performed by minimizing the cost or number of mismatched pairs in all columns in the MSA.

PAIRWISE SEQUENCE ALIGNMENT VERSUS MULTIPLE SEQUENCE ALIGNMENT

One of the important contributions of molecular biology to evolutionary analysis is the discovery that the DNA sequences of different organisms are often related. Similar genes are conserved across widely diverged species, often performing a similar or even identical function and at other times, mutating or rearranging to perform an altered function through the forces of natural selection. Thus many genes are represented in highly conserved forms in organisms. Through simultaneous alignment of the sequences of these genes, sequence patterns that have been subjected to alternation may be analysed. The potential for learning about the structure and function of molecules by multiple sequence alignment is so great and computational methods have received a great deal of attention. In MSA, sequences are aligned optimally by bringing the greatest number of similar characters into register in the same column of the alignment of more than two sequences that include matches, mismatches and gaps. This takes into account the degree of variation in all of the sequences and at the same time poses a very difficult challenge. The dynamic programming algorithm used for optimal sequence alignment of pairs of sequences can be extended to three sequences but for more than three sequences, only a small number of relatively short sequences may be characterized. Thus approximate methods are used, including progressive global alignment of the sequences starting with an alignment of the most alike sequences and then building an alignment by adding more sequences.[6]

Iterative methods make an initial alignment of a group of sequences and then revise the alignment to achieve a more reasonable result.

Alignment based on locally conserved patterns found in the same order in the sequences, use statistical and probabilistic models of the sequences.

A second computational challenge is identifying a reasonable method of obtaining a cumulative score for the "substitution in the column" of an MSA. Finally, the placement and scoring of gaps in various sequences of MSA presents an additional challenge.

The MSA of a set of sequences may also be viewed as an evolutionary history of the sequences. If these sequences in the MSA align very well, they are likely to be recently derived from a common ancestor sequence. Conversely a group of poorly aligned sequences share a more common complex and distant evolutionary relationship. The task of aligning a set of sequences, some more closely and other less closely related is by discovering the evolutionary relationship among the sequences.

The difficulty in aligning a group of sequences varies considerably with sequence similarity. On the one hand, if the amount of sequence variation is minimal, it is quite easy to align the sequences, even without the assistance of a computer program. On the other hand, if the amount of sequence variation is great, it may be difficult to find an optimal alignment of the sequences because so many combinations of substitutions, insertions and deletions, each predicting a different alignment, are possible.

Some of the many MSA programs are ClustalW or ClustalX; MSA; PRALINE; DIALIGN; MULTALIN, HMMER; MACAW, etc.

A group of similar sequences may define a protein family that may share a common biochemical function or evolutionary origin. Similar protein have been organized into databases of proteins families, which will be discussed later.

One application of MSA is in genome sequencing projects. Instead of cloning and arranging very large numbers of fragments of a large DNA molecule and then moving along the molecule and sequencing the fragments in order, random fragments of the large molecules are sequenced and those that overlap are found by the MSA program. This approach enables automated assembly of large sequences. Bacterial genome has been quite readily sequenced by this method.

Just as the alignment of a pair of nucleic acid or protein sequences can reveal whether or not there is an evolutionary relationship between the sequences, so can the alignment of three or more sequences reveal relationship among multiple sequences. If the structure of one or more members of the alignment is known, it may be possible to predict which amino acids occupy the same spatial relationship in other proteins in the alignment.

In nucleic acids, such alignments also reveal structural and functional relationships. For example, aligned promoters or a set of similarity-regulated genes may reveal consensus-binding sites for regulatory proteins.

Another use of consensus information retrieved from an MSA is for the prediction of specific probes for other members of the same group or family of similar sequences in the same or other organisms. There are both computer and molecular biology applications. Once a consensus pattern has been found, database searching programs may be used to find other sequences with a similar pattern. In the laboratory a reasonable consensus of such patterns may be used to define polymerase chain reaction (PCR) primers for amplification of related sequences.

Once MSA has been found, the number of types of changes in the aligned sequence residues may be used for a phylogenetic analysis. Each column in the alignment predicts the mutations that occurred at one site during evolution of the sequence family.

The close relationship between MSA and evolutionary tree construction shown is a short section of one MSA of four protein sequences including conserved and substituted positions, an insertion (of K) and a deletion (of L) below MSA shows an hypothetical evolutionary tree that could have generated these sequence changes. Each outer branch in the tree represents one of the sequences. The outer branches are also referred to as leaves. The deepest, oldest branch is that of sequence D, followed by A, then by B and C. The optimal alignment of several sequences can thereby be thought of as minimizing the number of mutational steps in an evolutionary tree for which the sequences are the outer branches or leaves. The mathematical solution to this problem was first outlined by Sankoff (1975). Fast MSA programs that are tree-based have since been developed. However, such an approach depends on knowing the evolutionary tree to perform an alignment. Often this is not the case. Usually, pairwise alignments are generated first and then used to predict the tree. In this example, the alignment can be explained by several different trees including the one shown in Figure 5.6.

Seq A	N	•	F	L	S
Seq A	N	•	F	•	S
Seq A	N	K	Y	L	S
Seq A	N	•	Y	L	S

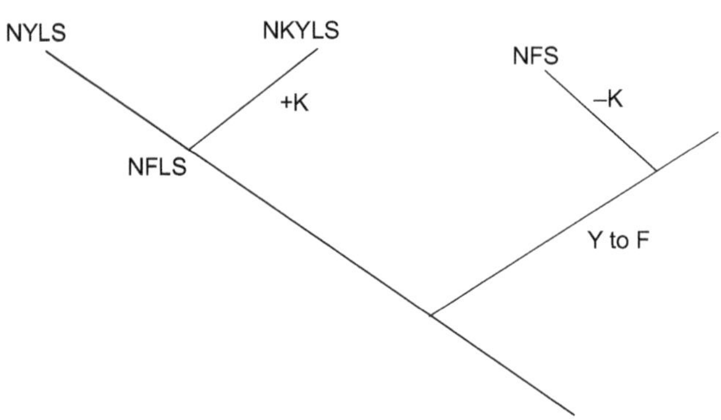

Figure 5.6 Tree diagram

Within the column are original characters that are present early, as well as other derived characters that appeared later in evolutionary time. In some cases the position is so important for function that mutational changes are not observed. In other cases the position is less important and substitutions are observed. Deletions and insertions may also be present in some regions of the alignment. Thus, starting with the alignment, one can hope to dissect the order of appearance of the sequences during evolution.

Clustal

Clustal performs a global multiple sequence alignment by a stepwise process. In step 1 it performs pairwise alignments of all the sequences provided by the user. In step 2 the scores obtained for the pairwise alignment are used to produce a phylogenetic tree and in step 3 the phylogenetic tree is used as a guide to align sequences sequentially. Thus the most closely related sequences are aligned first, and then additional sequences are added one by one to a profile of an existing MSA. The scoring of gaps is done in a manner different from that followed for a pairwise alignment.[7]

ClustalW calculates gaps in a novel way. The graphical version of ClustalW provides a versatile environment for doing MSA of sequences. Alignments can also be produced in a profile mode. Profile mode is typically used when MSA is already known for a set of sequences and when one wants to align a sequence of one MSA to another MSA. This is a very useful feature for finding conserved domains.

ClustalX has graphical user interface. Clustal was developed by Higgins and Sharp in 1988 and many improved versions were developed later. ClustalW is the recent version of Clustal with W standing for weighting to represent the ability of the program to provide weights to the sequence and program parameters.

Patterns in multiple sequence alignments Multiple sequence alignment is the process of aligning several related sequences, showing the conserved and unconserved residues across all of the sequences simultaneously. These conserved/unconserved residues form a pattern that can often be used to retrieve sequences that are distantly related to the original group of sequences. These distant relatives are extremely helpful in understanding the role that the groups of sequences play in the process of life.

Global multiple sequence alignments Global multiple sequence alignments are sequence alignments that require the participation of all sequence residues. A multiple sequence alignment shows the residue juxtaposition across the entire set of sequences, thus showing the conserved and unconserved residues across all of the sequences simultaneously.

Progressive pairwise approach The progressive pairwise approach relies on exhaustive pairwise alignments between all of the sequences to produce a measure of sequence relatedness. From this measure, an algorithm (UPGMA in Pileup, Neighbour Joining in ClustalW) is used to develop a joining order. This joining order corresponds to a tree that is used to proceed with the multiple sequence alignment. It should be noted that this tree is not an evolutionary tree. The tree is shown in Figure 5.7.

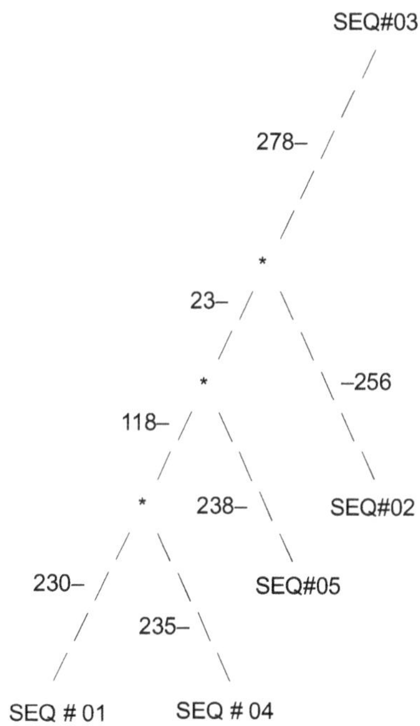

Figure 5.7 Tree of multiple sequence alignment (Progressive pairwise approach)

Five sequences have been taken for tree formation. seq 01, 04, have close relationship than seq 05 and 02 but a far relationship to seq 03.

After the joining order has been determined, sequences close to each other are aligned first. In the example above, SEQ#01 and SEQ#04 are the first two sequences to be aligned. The third sequence, SEQ#05, is then aligned with the two previously aligned sequences, SEQ#01 and SEQ#04. SEQ#02 is then aligned, followed by SEQ#03.

While this approach produces adequate results for many sets of sequences, the alignment produced by the procedure will vary depending on the joining order. Thus, joining the sequences in this order :

[SEQ#03 + SEQ#02] + [SEQ#05] + [SEQ#04] + [SEQ#01]

may not produce the same alignment as joining the sequences in the original order :

[SEQ#01 + SEQ#04] + [SEQ#05] + [SEQ#02] + [SEQ#03]

The advantages to this approach are that it requires only modest computer resources and that it is capable of aligning hundreds of sequences.

STORING ALIGNMENTS

Sequence alignments can be stored in a wide variety of text-based file formats. These were originally developed in conjunction with a specific alignment program or implementation. FASTA format and GenBank format are the web-based tools that allow a limited number of input and output formats. The output is not easily editable. Many programs are available for conversion—READSEQ or EMBOSS having a graphical interface or command line interface while several programming packages like BioPerl, BioRuby provide functions to do this.

REVIEW QUESTIONS

1. What are the various methods of aligning DNA and protein sequences?
2. Discuss BLAST and FASTA in detail emphasizing on the differences in methodologies.
3. Write an account on multiple sequence alignment.
4. How can alignments be represented and stored?
5. Elucidate dynamic programming.
6. Distinguish between local and global alignments using examples.
7. Write short notes on scoring matrices.

REFERENCES

1. Schneider, T.D. and Stephens, R.M. (1990). "Sequence logos: a new way to display consensus sequences." *Nucleic Acids Res.* 18: 6097–6100.
2. Brudno, M., Malde, S., Poliakov, A., Do, C.B., Couronne, O., Dubchak, I. and Batzoglou, S. (2003). "Global alignment: finding rearrangements during alignment." *Bioinformatics.* 19 Suppl. 1: i54–62.
3. Chothia, C. and Lesk, A.M. (1986). "The relation between the divergence of sequence and structure in proteins." *EMBO J.*5 (4): 823–6.
4. Thompson, J.D., Plewniak, F. and Poch, O. (1999). "A comprehensive comparison of multiple sequence alignment programs." *Nucleic Acids Res.* 27: 2682–90.

5. Mount, D.M. (2004). *Bioinformatics: Sequence and Genome Analysis*, 2nd edn. Cold Spring Harbor Laboratory Press: Cold Spring Harbor, NY. ISBN 0-87969-608-7.

6. Lipman, D.J, Altschul, S.F. and Kececioglu, J.D. (1989). "A tool for multiple sequence alignment." *Proc. Natl. Acad. Sci. USA* 86: 4412–5.

7. Higgins, D.G. and Sharp, P.M. (1988). "Clustal: a package for performing multiple sequence alignment on a microcomputer." *Gene.* 73 (1): 237–44.

6

PHYLOGENETIC ANALYSIS

Numberless intermediate varieties, linking closely together all the species of the same group.

Darwin 1859

OBJECTIVES

1. To compare more than two sequences to analyse the similarities and differences among them
2. To estimate evolutionary relationships among organisms

INTRODUCTION

Phylogenetics (Greek: *phyle/phylon*="tribe, race" and genetikos="relative to birth" from genesis "birth") is the study of evolutionary relatedness among various groups of organisms. Classification of organisms according to similarity, has been richly complemented by phylogenetics but remains methodologically and logically distinct.[1]

Phylogenetics attempts to reconstruct evolutionary history. In a phylogenetic analysis, one compares the results of evolutionary processes, be it the shape and size of specific bones or patterns of DNA or protein sequences, in an attempt to determine how different groups or species may have been derived during evolution. Looking at the molecular sequences can give insight into the nature of the processes, which lead to divergence, in addition to analysing the relative degree of relatedness.

The Origin of Molecular Phylogenetics

Macromolecular data, meaning gene (DNA) and protein sequences, are accumulating at an increasing rate because of recent advances in molecular biology. For the evolutionary biologist, the rapid accumulation of sequence data from whole genomes has been a major advance, because the very nature of DNA allows it to be used as a document of evolutionary history. Comparison of the DNA sequences of various genes between different organisms can tell a scientist a lot about the relationships of organisms that cannot otherwise be inferred from morphology or an organism's outer form and inner structure. Because genomes evolve by the gradual accumulation of mutations, the amount of nucleotide sequence difference between a pair of genomes from different organisms should indicate how recently those two genomes shared a common ancestor. Two genomes that diverged in the recent past should have fewer differences than two genomes whose common ancestor is more ancient.

Therefore, by comparing different genomes with each other, it should be possible to derive evolutionary relationships between them, which is the major objective of molecular phylogenetics. Molecular phylogenetics attempts to determine the rates and patterns of change occurring in DNA and proteins and to reconstruct the evolutionary history of genes and organisms. Two general approaches may be taken to obtain this information. In the first approach, scientists use DNA to study the evolution of an organism. In the second approach, different organisms are used to study the evolution of DNA. Point mutation in DNA leads to variation at the molecular level and thus gives rise to new species in the course of evolution.

Some variations include:

1. Random mutation, which can be seen as random genetic drift in the absence of selective pressure.
2. Sequence duplication, which may be duplication of small segments, genes, or even whole genomes.
3. Recombination, which includes transposons, translocations and viral activity, to mix up

sequences within an organism, to remove sequences or to introduce sequences from another organism.

MOLECULAR PHYLOGENETIC ANALYSIS

Fundamental elements like the macromolecules, especially gene and protein sequences, have surpassed morphological and other organismal characters as the most popular forms of data for phylogenetic analysis. Phylogenetic tree-building models presume particular evolutionary models. For any given set of data, these models may be violated because of various occurrences, such as the transfer of genetic material between organisms.

A straightforward phylogenetic analysis consists of four steps:

1. Alignment-building
2. Determining sequence variation
3. Tree building
4. Tree evaluation

MOLECULAR CLOCK HYPOTHESIS

This hypothesis states that nucleotide substitutions or amino acid substitutions in proteins that are being compared, occur at a constant rate, that is, the degree of difference between the two sequences can be used to assign a date at which their ancestral sequence diverged. The rate of molecular change differs among groups of organisms among genes, and even among different parts of the same gene. Furthermore, molecular clocks require calibration with fossils to determine timing of origin of clades and thus their accuracy is crucially dependent on the fossil record, or lack thereof, for the groups under study. Fossil DNA older than about 25,000–50,000 years is virtually empty of phylogenetic signals except in rare instances and therefore traditional morphological studies of extinct and extant organisms remain a crucial component of phylogenetic analysis.

THE IMPORTANCE OF MOLECULAR PHYLOGENETICS

The field of molecular phylogenetics has grown, both in size and in importance, since its inception in the early 1990s, attributable mostly to advances in molecular biology and more rigorous methods for phylogenetic tree building. The importance of phylogenetics has also been greatly enhanced by the successful application of tree reconstruction, as well as other phylogenetic techniques. Broadly speaking, the relationships established by phylogenetic trees often describe the evolutionary history of a species and hence, its phylogeny; the historical relationships among lineages or organisms or their parts, such as their genes. Phylogenies may be thought of as a natural and meaningful way to order data, with an enormous amount of evolutionary information contained within their branches. Scientists working in these different areas can then use these phylogenies to study and elucidate the biological processes occurring at many levels of life's hierarchy.

METHODS OF PHYLOGENETIC ANALYSIS

There is considerable debate regarding the best approach to take when analysing sequence alignments for phylogenetic relationships. Accepted approaches include distance calculations, parsimony and maximum likelihood. Nowadays Bayesian analysis is gaining popularity.

The associated concepts in phylogenetic analysis are

- Phylogenetics vs. taxonomy[2]
- Cladistic vs. phenetic
- Clustering
- Parsimony vs. maximum likelihood

Two major groups of analyses exist to examine phylogenetic relationships: phenetic methods and cladistic methods. It is important to note that phenetics and cladistics have had an uneasy relationship over the last 40 years or so. Most of today's evolutionary biologists favour cladistics, although a strictly cladistic approach may result in counter-intuitive results.

Phenetic Method of Analysis

Phenetics, also known as numerical taxonomy, involves the use of various measures of overall similarity for the ranking of species. There is no restriction on the number or type of characters (data) that can be used, although all data must be first converted to a numerical value, without any character weighting. Each organism is then compared with every other for all characters measured and the number of similarities (or differences) is calculated. The organisms are then clustered in such a way that the most similar ones are grouped close together and the more different ones are linked more distantly. The taxonomic clusters called phenograms, that result from such an analysis do not necessarily reflect genetic similarity or evolutionary relatedness. The lack of evolutionary significance in phenetics has meant that this system has had little impact on animal classification and as a consequence, use of phenetics has been declining in recent years.

Phenetics failed to create truly evolutionary groups because organisms can be similar because they live in similar environments or because they make their living in similar ways, not just because they are descended from a common ancestor. Despite the failure of phenetics to reconstruct evolutionary relationships, the numerical methods themselves have applications to a variety of other problems, such as delimiting variation within and between populations of individuals. This approach can be particularly useful to circumscribe species with associated ranges of variation from collections of individual specimens.

Cladistic Method of Analysis

An alternative approach to diagrammatic relationships between taxa is called cladistics. The basic assumption behind cladistics is that members of a group share a common evolutionary history. Thus, they are more closely related to one another than they are to other groups of organisms. Related groups of organisms are recognized because they share a set of unique features (apomorphies) that were not present in distant ancestors but which are shared by most or all of the organisms within the group. An apomorphy is a derived character. Apomorphies that are shared between sister taxa are called synapomorphies. Therefore, in contrast to phenetics, cladistics groupings do not depend on whether organisms share physical traits but depend on their evolutionary relationships. Indeed, in cladistic analysis two organisms may share numerous characteristics but still be considered members of different groups. Analysis entails a number of assumptions.[3]

For example, species are assumed to arise primarily by bifurcation or separation of the ancestral lineage; species are often considered to become extinct upon hybridization (cross-breeding); and hybridization is assumed to be rare or absent. In addition, cladistic groupings must possess the following characteristics: all species in a grouping must share a common ancestor and all species derived from a common ancestor must be included in the taxon. The most important insight of cladistics is that if one takes all character states shared by a number of organisms into account (i.e., overall similarity), one will not necessarily get a classification that reflects actual evolutionary relationships. Instead, one needs to concentrate only on certain characters—those that provide evolutionary information.

PHYLOGENETIC TREES

Systematics describes the pattern of relationships among taxa and is intended to help us understand the history of all life, but history is not something we can see. It has happened once and leaves only clues to the actual events. Scientists use these clues to build hypotheses or models of life's history. In phylogenetic studies, the most convenient way of visually presenting evolutionary relationships among a group of organisms is through illustrations called phylogenetic trees.

The divergence mode of evolution has the following leads:

1. *Node*—represents a taxonomic unit. This can be either an existing species or an ancestor.
2. *Branch*—defines the relationship between the taxa in terms of descent and ancestry.
3. *Branch length*—represents the number of changes that have occurred in the branch.
4. *Topology*—the branching patterns of the tree.
5. *Root*—the common ancestor of all taxa.
6. *Distance scale*—represents the number of differences between organisms or sequences.
7. *Clade*—a group of two or more taxa or DNA sequences that includes both their common ancestor and all of their descendents.
8. *Operational taxonomic unit (OTU)*—taxonomic level of sampling selected by the user to be used in a study, such as individuals, populations, species, genera or bacterial strains.

A phylogenetic tree is composed of nodes, each representing a taxonomic unit (species, populations, individuals) and branches, which define the relationship between the taxonomic units in terms of descent and ancestry. Only one branch can connect any two adjacent nodes. The branching pattern of the tree is called the topology and the branch length usually represents the number of changes that have occurred in the branch. This is called a scaled branch. Scaled trees are often calibrated to represent the passage of time. Such trees have a theoretical basis in a particular gene or genes under analysis. Branches can also be unsealed, which means that the branch length is not proportional to the number of changes that have occurred, although the actual number may be indicated numerically somewhere on the branch. Phylogenetic trees may also be either rooted or unrooted. In rooted trees, there is a particular node, called the root, representing a common ancestor, from which a unique path leads to any other node. An unrooted tree only specifies the relationship among species, without identifying a common ancestor or evolutionary path.

A typical gene-based phylogenetic tree is depicted in Figure 6.1. This tree shows the relationship among the four homologous genes A, B, C and D. The topology of this tree consists of four external nodes (A, B, C and D), each one representing one of the four genes, and two internal nodes (e and f) representing ancestral genes. The branch lengths indicate the degree of evolutionary differences between the genes. This particular tree is unrooted, it is only an illustration of the relationships between the genes A, B, C, and D and does not signify anything about the series of evolutionary events that led to these genes. A noded tree is often referred to as an inferred tree. This is to emphasize that this type of illustration depicts only the series of evolutionary events that are inferred from the data under study and may not be the same as the true tree or the tree that depicts the actual series of evolutionary events that occurred.

Figure 6.1 Gene-based phylogenetic tree

Evolutionary Trees

An evolutionary tree is a two-dimensional graph showing the evolutionary relationship among a set of items being compared. This set can be organisms, genes or DNA sequences. Each unit in the set is referred to as a taxon. Each taxon will be defined by a distinct unit on the tree.

An evolutionary tree is composed of outer branches or leaves that represent the taxa, and nodes and branches representing the relationships among the taxa. Two taxa that are derived from the same common ancestor will share a node in the graph. In general, approaches to designing evolutionary trees are made to define the length of each branch to the next node according to the number of sequence-level changes that occurred.

Rooted trees In a rooted tree topology, one sequence (the root) is defined to be the common ancestor of all of the other sequences. A unique path leads from the root node to any other node and the direction of the path indicates evolutionary time. The root is chosen by including a sequence from an organism that is thought to have branched off earlier than the other sequences. When the number of sequences increases, the number of possible rooted trees increases very rapidly. In some cases, a bifurcating binary tree is the best model to simulate evolutionary events in which case one species branches off into two separate species (Figure 6.2).

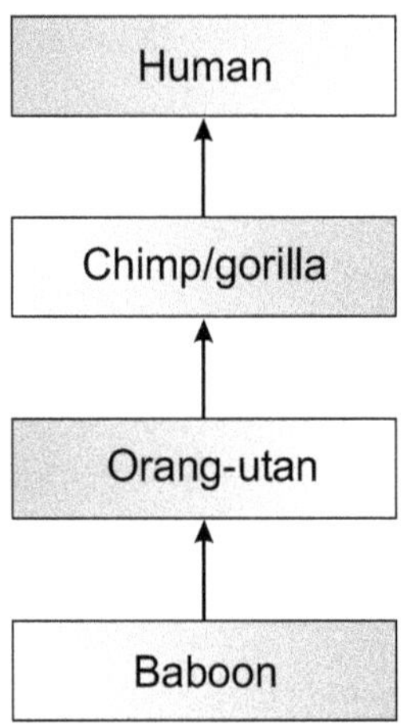

Figure 6.2 Example of a rooted tree

Unrooted trees (star topology) An unrooted tree (sometimes referred to as a star topology) shows the evolutionary relationship among sequences, without revealing the location of the oldest ancestry (Figure 6.3). There are fewer choices for an unrooted tree than a rooted tree.

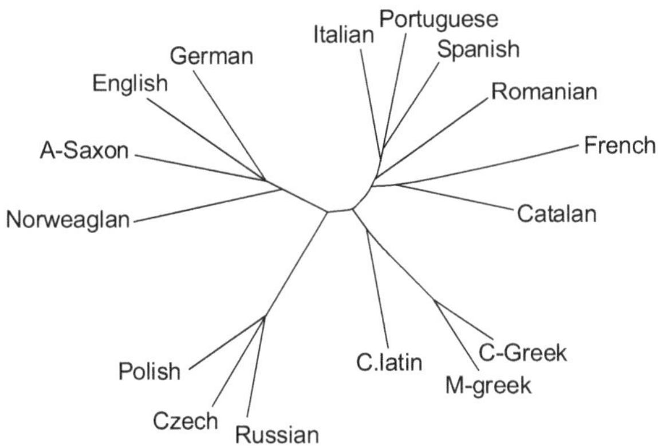

Figure 6.3 Example of an unrooted tree

METHODS FOR DETERMINING EVOLUTIONARY TREES

There are three methods used to calculate the tree(s) that best accounts for the observed variation in a set of sequences. These methods are maximum parsimony, distance and maximum likelihood.

Maximum Parsimony

Maximum parsimony methods predict the evolutionary tree that minimizes the number of steps required to generate the observed variation in the sequences. In order to construct a tree using maximum parsimony, a multiple sequence alignment must first be obtained. For each aligned position, phylogenetic trees that require the smallest number of evolutionary changes to produce the observed sequence changes are identified. This continues for each position in the alignment. Those trees that produce the smallest number of changes overall for all sequence positions are identified. This is a rather time-consuming algorithm that works well only if the sequences have a strong sequence similarity.

```
    5 7 9
1 A A GAG T G C A
2 A G CCG T G C G
3 A G ATA T G C A
4 A G AGA T C C G
```

Figure 6.4 Multiple alignment for phylogeny

Consider the example in Figure 6.4. There are a total of four sequences, which gives a possibility of three different unrooted trees as shown in Figure 6.5. In this case some sites are informative and other sites are not. An informative site has the same sequence character in at least two different sequences. Only the informative sites need to be considered.

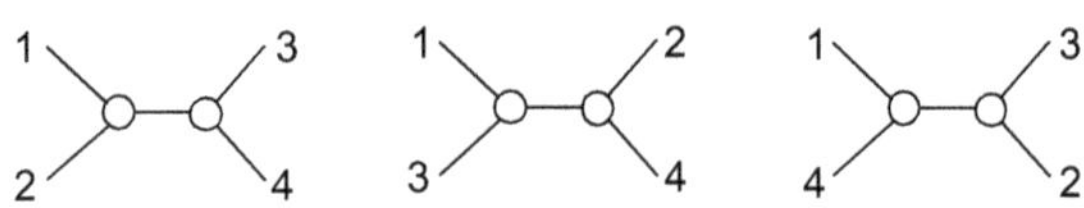

Figure 6.5 Possible trees from four sequences

In this case, adding the number of changes at each informative site for each tree and picking the tree requiring the least total number of changes, obtain the optimal tree.

For a large number of sequences the number of trees to examine becomes so large that it might not be possible to examine all possible trees. Some programs, such as PAUP (Phylogenetic analysis using parsimony) add features that will allow the user to invoke a heuristic method that will keep representative trees that best fit the data.

The informative sites in the example alignment are positions 5, 7, and 9 in Figure 6.4. The possible trees and the number of rearrangements for each in the informative sites are to be figured out.

One problem with determining evolutionary distance between sequences is that columns representing greater variation dominate the analysis. In order to overcome this problem of determining long branch lengths is to look only at transversion events, which are the most significant base changes (i.e., changes a purine to a pyrimidine or vice versa). This is referred to as Lake's method of invariants.

Distance Methods

The distance method for construction of phylogenetic trees looks at the number of changes between each pair in a group of sequences to produce a phylogenetic tree of the group. The goal of distance methods is to identify a tree that positions neighbours correctly and that also has branch lengths, which reproduce the original data as closely as possible.

ClustalW This uses the neighbour-joining method as a guide to multiple sequence alignments. The PHYLIP suite of programs employs neighbour-joining methods, e.g. distance analysis programs in PHYLIP.

FITCH This estimates a phylogenetic tree assuming additivity of branch lengths using the Fitch–Margoliash method.

KITSH This is the same as FITCH, but under the assumption of a molecular clock.

NEIGHBOUR This estimates phylogenies using the neighbour-joining (no molecular clock assumed) or unweighted pair group method with arithmetic mean (UPGMA) (molecular clock assumed).

For phylogenetic analysis, the distance score is counted as either the number of mismatched positions in the alignment or the number of sequence positions that must be changed to generate the other sequence used.

The success of distance method depends on the degree to which the distances among a set of sequences can be made additive on a predicted evolutionary tree.

Consider the alignment:

```
A  ACGCGTTGGGCGATGGCAAC
B  ACGCGTTGGGCGACGGTAAT
C  ACGCATTGAATGATGATAAT
D  ACACATTGAGTGATAATAAT
```

The distances between these sequences can be shown as a table as shown in Figure 6.6.

	A	B	C	D
A	–	3	7	8
B	–	–	6	7
C	–	–	–	3
D	–	–	–	–

Figure 6.6 Distance table

Note Distances are nothing but the number of changes in bases/amino acids between any two sequences under comparison.

Using this information, an unrooted tree showing the relationship between these sequences can be drawn as shown in Figure 6.7.

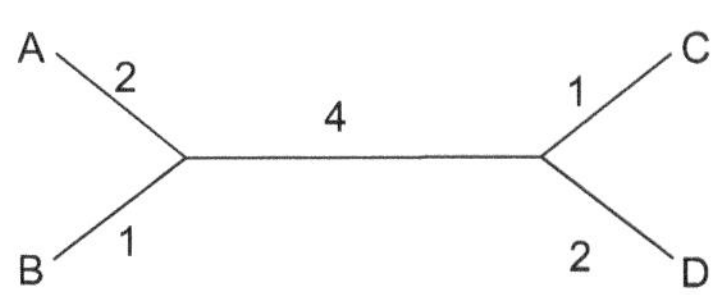

Figure 6.7 Unrooted tree resulting from the above four sequences

Fitch and Margoliash Method

The Fitch and Margoliash method uses a distance table. The sequences are combined in threes to define the branches of the predicted tree and to calculate the branch lengths of the tree. Some features of these trees include the following.

1. They find the most closely related pairs of sequences (A, B).

2. They treat the rest of the sequences as a composite. Calculate the average distance from A to all others; and from B to all others.
3. These trees use these values to calculate the length of the edges A and B.
4. They treat A and B as a composite. Calculate the average distances between AB and each of the other sequences. Create a new distance table.
5. They identify next pair of related sequences and begin with step 1.
6. They subtract extended branch lengths to calculate lengths of intermediate branches.
7. They repeat the entire process with all possible pairs of sequences.
8. They calculate predicted distances between each pair of sequences for each tree to find the best tree.

Neighbour-Joining Algorithm

The neighbour-joining method is very similar to the Fitch–Margoliash method. The sequences that should be joined are chosen to give the best least-square estimates of the branch lengths that most closely reflect the actual distances between the sequences.

The neighbour-joining method begins by creating a star topology in which neighbours are joined (Figure 6.8).

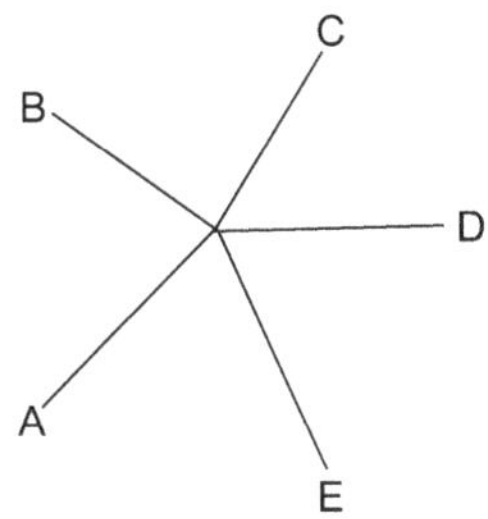

Figure 6.8 Star topology of neighbour-joining method

The tree is modified by joining pairs of sequences. The pair to be joined is chosen by calculating the sum of the branch lengths for the corresponding tree. The sum of the branch lengths is calculated as follows:

The pair that results in the smallest branch length is then chosen to be the pair that is joined. Based on this choice, the Fitch–Margoliash algorithm is used to compute the actual branch lengths. After the pair has been joined, a new distance table is created with the recently joined sequences now entered as a composite. The neighbour-joining algorithm chooses the next pair of sequences to join and the F–M algorithm computes the branch lengths.

The process continues until the correctly branched tree and distances have been identified.

Unweighted Pair Group Method with Arithmetic Mean (UPGMA)

It works by clustering the sequences, starting with more similar sequences and working towards more distant sequences.

The process assembles a tree upwards, with each node being added above the others and the edge lengths being determined by the difference in the heights of the nodes.

DIFFICULTIES WITH PHYLOGENETIC ANALYSIS

Phylogenetic analysis would be easier if evolution occurred in a vertical fashion. However, horizontal or lateral transfer of genetic material occurs, which makes it difficult to determine the phylogenetic origin of some evolutionary events.

If a gene is under selective pressure in different organisms, it can be rapidly evolving. Such an evolution can mask earlier changes that had occurred phylogenetically. In addition, different regions of a genome are under different pressures and therefore different sites within two comparative sequences may be evolving at different rates.

Rearrangements of genetic material can also lead to false conclusions with phylogenetic analysis, especially if two sequences of different evolutionary origins are placed next to each other.

Gene duplication events also cause problems with phylogenetic analysis, since the duplicated genes can evolve along separate pathways, leading to different functions.

PHYLOGENETIC SOFTWARE RESOURCES

General-purpose Packages

- PHYLIP
- PAUP*
- MEGA
- Phylo_win
- ARB
- DAMBE
- PAL
- Bionumerics
- Mesquite
- PaupUp
- BIRCH
- Bosque
- EMBOSS
- phangorn

Parsimony Programs

- PHYLIP
- PAUP*
- Hennig86
- MEGA
- RA
- Nona
- PHYLIP
- CAFCA
- Phylo_win
- sog
- gmaes
- LVB
- GeneTree
- TAAR
- ARB
- DAMBE
- MALIGN
- POY
- Gambit
- TNT
- GelCompar II
- Bionumerics
- Network
- TCS
- GAPars
- CRANN
- Mesquite
- PAST
- FootPrinter
- BPAnalysis
- Simplot
- Parsimov
- NimbleTree
- PaupUp
- Notung
- galaxie
- BIRCH
- IDEA
- PSODA
- PRAP
- SeqState
- Bosque
- PhyloNet
- EMBOSS
- phangorn

Distance Matrix Methods

- PHYLIP
- PAUP*
- MEGA
- MacT
- ODEN
- TREECON
- DISPAN
- RESTSITE
- NTSYSpc
- METREE
- TreeTree
- GDA
- SeqPup
- PHYLTEST
- Lintre
- WET
- Phylo_win
- POPTREE
- Gambit
- gmaes
- DENDRON
- BIONJ
- TFPGA
- MVSP
- ARB
- Darwin
- T-REX
- sendbs
- nneighbor
- DAMBE
- weighbor
- DNASIS
- MINSPNET
- PAL
- Arlequin
- vCEBL
- HY-PHY
- Vanilla
- GelCompar II
- Bionumerics
- qclust
- TCS
- Populations
- Winboot
- SYN-TAX
- PTP
- SplitsTree
- FastME
- APE
- MacVector
- QuickTree
- Simplot
- ProfDist
- START
- STC
- NimbleTree
- CBCAnalyzer
- PaupUp
- galaxie
- Geneious
- BIRCH
- SEMPHY
- FASTML
- Rate4Site
- SWORDS
- IDEA
- FAMD
- Bosque
- GAME
- TreeFit
- EMBOSS
- phangorn
- FastTree
- Hadtree, Prepare and Trees

- Fingerprinting II Informatix Software
- Discovery Studio Gene
- Bioinformatics_Toolbox

Computation of Distances

- PHYLIP
- PAUP*
- RAPDistance
- MULTICOMP
- Microsat
- DIPLOMO
- OSA
- DISPAN
- RESTSITE
- NTSYSpc
- GCUA
- DERANGE2
- POPGENE
- TFPGA
- REAP
- MVSP
- RSTCALC
- Genetix
- DISTANCE
- Darwin
- sendbs
- K2WuLi
- GeneStrut
- Arlequin
- DAMBE
- DnaSP
- PAML
- puzzleboot
- PAL
- Vanilla
- GelCompar II
- Bionumerics
- qclust
- Populations
- Winboot
- FSTAT
- SYN-TAX
- Phylo_win
- Phyltools
- MSA
- APE
- YCDMA
- NSA
- T-REX
- LDDist
- DIVAGE
- Genepop
- START
- Swaap
- Swaap PH
- GeneContent
- SPAGeDi
- PaupUp
- SEMPHY
- SWORDS
- FAMD
- GAME
- GenoDive
- analysis
- TreeFit
- EMBOSS
- rRNA phylogeny

- CBCAnalyzer
- TREE-PUZZLE
- Bioinformatics_Toolbox
- Hadtree, Prepare and Trees

Maximum Likelihood and Bayesian Methods

- PHYLIP
- PAUP*
- fastDNAml
- MOLPHY
- PAML
- Spectrum
- SplitsTree
- PLATO
- SeqPup
- Phylo_win
- PASSML
- ARB
- Darwin
- BAMBE
- DAMBE
- Modeltest
- TreeCons
- PAL
- dnarates
- TrExMl
- HY-PHY
- Vanilla
- DT-ModSel
- Bionumerics
- RevDNArates
- MrBayes
- rate-evolution
- CONSEL
- EDIBLE
- Mesquite
- PTP
- Treefinder
- MetaPIGA
- RAxML
- PHASE
- PHYML
- BEAST
- r8s-bootstrap
- MrMTgui
- MrModeltest
- BootPHYML
- p4
- Porn*
- SIMMAP
- Spectronet
- Rhino
- IM
- ProtTest
- Simplot
- MDIV
- MrAIC
- Modelfit
- IQPNNI
- PARAT
- ALIFRITZ
- PhyNav
- DPRML
- MultiPhyl
- NimbleTree
- PaupUp
- SSA
- BAli-Phy
- CoMET

- BIRCH
- GARLI
- PluginSEMPHY
- aLRT
- BEST
- PROCOV
- AMBIORE
- Leaphy
- rRNA phylogeny
- PHYLLAB
- EMBOSS
- TREE-PUZZLE
- ModelGenerator
- VeryfastDNAml
- MrBayes tree scanners
- Hadtree, Prepare and Trees
- Mac5
- PHYSIG
- FASTML
- McRate
- EREM
- DART
- PRAP
- NHML
- Bosque
- NEPAL
- phangorn
- fastDNAmlRev
- BayesPhylogenies
- MrBayesPlugin
- Kakusan2
- MrBayes
- Rate4Site
- PhyloBayes
- IDEA
- PhyloCoCo
- SeqState
- SLR
- Concaterpillar
- CodeAxe
- bms_runner

Quartets Methods

- TREE-PUZZLE
- PHYLTEST
- PICA
- PhyloQuart
- Gambit
- STC
- LEVEL2
- SplitsTree
- GEOMETRY
- Darwin
- Willson quartets programs
- IQPNNI
- Quartet Suite
- Bosque

Artificial-intelligence and Genetic Algorithms Methods

- PTP
- MetaPIGA
- GAPars
- GARLI

Invariants (or Evolutionary Parsimony) Methods

- PHYLIP
- PaupUp
- EMBOSS
- PAUP*
- BIRCH

Interactive Tree Manipulation

- MacClade
- TreeTool
- TreeEdit
- RadCon
- EDIBLE
- TreeView
- Notung
- BIRCH
- Dendroscope
- TreeToy
- PhyloWidget
- Bioinformatics_Toolbox
- PHYLIP
- ARB
- TreeExplorer
- Mavric
- Mesquite
- TreeMe
- TreeDyn
- SimpleClade
- Forest
- Phyutility
- TreeJuxtaposer
- PDAP
- WINCLADA
- TreeThief
- T-REX
- Treefinder
- ArboDraw
- TreeMaker
- TreeIllustrator
- Phylocom
- EMBOSS

Looking for Hybridization or Recombination Events

- PLATO
- TOPALi
- Network
- T-REX
- PIST
- START
- cBrother
- HGT
- Concaterpillar
- reticulate
- partimatrix
- TCS
- PPH
- IM
- Likewind
- TreeMos
- PhyloNet
- Codeml3X
- RecPars
- LARD
- SiScan
- Spectronet
- Simplot
- DualBrothers
- EEEP
- EvolSimulator
- RDP3

Bootstrapping and other Measures of Support

- PHYLIP
- TreeRot
- DISPAN
- Lintre
- MEGA
- TAXEQ3
- DAMBE
- TrExMl
- MrBayes
- LVB
- Mesquite
- Treefinder
- PHASE
- r8s-bootstrap
- PAUP*
- DNA Stacks
- TreeTree
- sog
- PICA
- TreeCons
- puzzleboot
- PAL
- CONSEL
- EDIBLE
- Phylo_win
- RAxML
- PHYML
- T-REX
- AutoDecay
- OSA
- PHYLTEST
- POPTREE
- ModelTest
- BAMBE
- Gambit
- PHYCON
- Populations
- Winboot
- PAST
- Phyltools
- BEAST
- MrMTgui

- MrModeltest
- BootPHYML
- Porn*
- ProtTest
- Simplot
- MCS
- Permute!
- ELW
- MultiPhyl
- GHOSTS
- PaupUp
- Geneious
- BIRCH
- scaleboot
- aLRT
- PhyloBayes
- SWORDS
- CTree
- PRAP
- FAMD
- bootscore
- PhyloNet
- PHYLLAB
- Phyutility
- EMBOSS
- FastTree
- Bionumerics
- CodonBootstrap
- ModelGenerator
- Random Cladistics
- BayesPhylogenies
- Discovery Studio Gene
- MrBayes tree scanners

Compatibility Analysis

- COMPROB
- PHYLIP
- PICA
- reticulate
- partimatrix
- CLINCH
- PPH
- Spectronet
- BIRCH
- EMBOSS

Consensus Trees, Subtrees, Supertrees, and Distances Between Trees

- COMPONENT
- TREEMAP
- NTSYSpc
- PHYLIP
- PAUP*
- REDCON

- TAXEQ3
- RadCon
- Treefinder
- SuperTree
- BIRCH
- Rainbow
- PhyloSort
- PhyloNet
- phangorn
- Supertree scripts
- TreeCons
- Mesquite
- Clann
- PaupUp
- TopD/fMts
- PhySIC
- FAMD
- Phyutility
- HeuristicMRF2
- Robinson and Foulds distance
- MEGA
- PAST
- PhyNav
- Supertree
- Quartet Suite
- EEEP
- bootscore
- EMBOSS

Tree-based Sequence Alignment

- TreeAlign
- GeneDoc
- DAMBE
- DNASIS
- T-Coffee
- galaxie
- MAFFT
- MUSCLE
- ClustalW
- TAAR
- POY
- FootPrinter
- ArboDraw
- Geneious
- SATCHMO
- Bosque
- MALIGN
- Ctree
- ALIGN
- ALIFRITZ
- BAli-Phy
- BIRCH
- DART
- EMBOSS

Gene Duplication and Genomic Analysis

- DERANGE2
- BPAnalysis
- TopD/fMts
- gtp
- Tree Tracker
- Mgenome
- Notung
- DTscore
- DTdraw
- bms_runner
- FORESTER
- GeneContent
- DART
- EvolSimulator

Biogeographic Analysis and Host-parasite Comparison

- COMPONENT
- DIVA
- TREEMAP
- TreeFitter
- MDIV
- GEODIS
- Tarzan
- ParaFit
- Phylocom
- AxParafit

Comparative Method Analysis

- PHYLIP
- CAIC
- COMPARE
- CMAP
- CoSta
- PDAP
- ACAP
- ANCML
- RIND
- MacroCAIC
- MacClade
- Mesquite
- APE
- Jevtrace
- SIMMAP
- PHYLOGR
- TreeSAAP
- Permute!
- Parsimov
- DIVERGE
- IDC
- OUCH
- BIRCH
- BayesTraits
- PHYSIG
- Cactus-Pie
- Phylocom
- pcca
- EMBOSS
- bms_runner
- Phylogenetic Independence

Simulation of Trees or Data

- Bi-De
- Seq-Gen
- PSeq-Gen
- COMPARE
- ROSE
- PAML
- ProSeq
- PAL
- Vanilla
- acClade
- EDIBLE
- Mesquite
- Treefinder
- Network
- Hetero

- Phyl-O-Gen
- MESA
- SGRunner
- apTreeshape
- Simprot
- EREM
- DAWG
- indel-Seq-Gen
- COMPONENT
- EvolveAGene3
- EvolSimulator
- Treevolve and PTreevolve

Examination of Shapes of Trees

- MacroCAIC
- Genie
- PAL
- Vanilla
- RadCon
- APE
- Tracer
- TreeScan
- MESA
- apTreeshape
- TreeStat
- CTree
- Phyutility
- laser
- SymmeTREE
- BRANCHLENGTH

Clocks, Dating and Stratigraphy

- PHYLIP
- QDate
- Diversi
- K2WuLi
- Modeltest
- PAML
- TipDate
- RRTree
- vCEBL
- TreeEdit
- HY-PHY
- MEGA
- PAL
- r8s
- PAST
- Treefinder
- Network
- APE
- BEAST
- MrMTgui
- MrModeltest
- SymmeTREE
- Porn*
- Rhino
- GHOSTS
- Cadence
- Multidivtime
- CodonRates
- BIRCH
- Brownie
- PATHd8
- McRate
- PhyloBayes

- ⊙ Cactus-Pie
- ⊙ GRate
- ⊙ NHML
- ⊙ TreeFit
- ⊙ rate-evolution
- ⊙ Ultrametric Check
- ⊙ BRANCHLENGTH
- ⊙ MULTIDIVTIME HELPER

Model Selection

- ⊙ Modeltest
- ⊙ MrMTgui
- ⊙ MrModeltest
- ⊙ Porn*
- ⊙ ProtTest
- ⊙ MrAIC
- ⊙ Modelfit
- ⊙ DT-ModSel
- ⊙ BayesTraits
- ⊙ Kakusan2
- ⊙ MAPPS
- ⊙ DART
- ⊙ Concaterpillar
- ⊙ ModelGenerator

Description or Prediction of Data From Trees

- ⊙ CONSERVE
- ⊙ EDIBLE
- ⊙ Jevtrace
- ⊙ TreeSAAP
- ⊙ MESA
- ⊙ RPT
- ⊙ TreeDis
- ⊙ RIND
- ⊙ SIMMAP
- ⊙ DIVERGE
- ⊙ FASTML

Tree Plotting/Drawing

- ⊙ PHYLIP
- ⊙ PAUP*
- ⊙ TreeTool
- ⊙ TreeView
- ⊙ NJplot
- ⊙ DendroMaker
- ⊙ Tree Draw Deck
- ⊙ ARB
- ⊙ unrooted
- ⊙ DAMBE
- ⊙ TREECON
- ⊙ Mavric

- TreeExplorer
- TreeThief
- Bionumerics
- FORESTER
- MacClade
- MEGA
- Mesquite
- APE
- T-REX
- Spectronet
- TreeSetViz
- TreeGraph
- ArboDraw
- PaupUp
- Notung
- TreeDyn
- DigTree
- Geneious
- BIRCH
- Paloverde
- MrEnt
- FigTree
- HyperTree
- TreeIllustrator
- Dendroscope
- CTree
- RPT
- TreeToy
- TreeSnatcher
- DTdraw
- PHYLLAB
- EMBOSS
- PhyloWidget
- Phylodendron
- TreeJuxtaposer
- GeoPhyloBuilder
- Bioinformatics_Toolbox
- Phylogenetic Tree Drawing

Sequence Management/Job Submission

- GDE
- MUST 2000
- DNA Stacks
- SeqPup
- ARB
- DAMBE
- BioEdit
- Bionumerics
- W2H
- Phyledit
- Simplot
- DPRML
- NimbleTree
- galaxie
- Geneious
- BIRCH
- TOPALi
- MBEToolbox
- PISE
- Bosque
- EMBOSS
- GeneStudio Pro
- Random Cladistics
- Bioinformatics_Toolbox
- Singapore PHYLIP web interface

Teaching about Phylogenies

- Phylap
- Dnatree
- SimpleClade
- Phylogenetic Investigator

SOME PHYLOGENETIC TERMINOLOGY

- A monophyletic grouping is one in which all species share a common ancestor and all species derived from that common ancestor are included. This is the only form of grouping accepted as valid by cladists.
- A paraphyletic grouping is one in which all species share a common ancestor but not all species derived from that common ancestor are included.
- A polyphyletic grouping is one in which species that do not share an immediate common ancestor are lumped together, while excluding other members that would link them.
- Homologs are most commonly defined as orthologs, paralogs or xenologs.
- Orthologs are homologs produced by speciation. They represent genes derived from a common ancestor that diverged because of divergence of the organism. Orthologs tend to have similar function.
- Paralogs are homologs produced by gene duplication and represent genes derived from a common ancestral gene that duplicated within an organism and then diverged. Paralogs tend to have different functions.

- Xenologs are homologs resulting from the horizontal transfer of a gene between two organisms. The function of xenologs can be available, depending on how significant the change in context was for the horizontally moving gene.
- Clades is a group of monophyletic DNA sequences that make up all of the sequences included in the analysis that have descended from a particular common ancestral sequence.
- Parsimony is an approach that decides between different tree topologies by identifying the one that involves the shortest evolutionary pathway. This is the pathway that requires the smallest number of nucleotide changes to go from the ancestral sequence, at the root of the tree, to all of the present-day sequences that have been compared.

REVIEW QUESTIONS

1. Describe the phylogenetic trees in detail.
2. Distinguish between rooted and unrooted trees.
3. Write short notes on each of the following:
   i. Maximum likelihood
   ii. Maximum parsimony
   iii. Distance
   iv. Neighbour-joining method
   v. Fitch–Margoliash method

REFERENCES

1. Edwards, A.W.F. and Cavalli-Sforza, L.L. (1964). "Phenetic and phylogenetic classification: Reconstruction of evolutionary trees." *Systematics Assoc. Publ.* 6: 67–76.
2. Queiroz, K. and Gauthier, J. (1992). "Phylogenetic taxonomy." *Annual Review of Ecology and Systematics.* Volume 23, pp. 449–480.
3. Arnold, G. Kluge and Alan, J. Wolf. (1993). "Cladistics: What's in a word?" *Cladistics.* 9(2):183–199.

7

AVENUES OF BIOINFORMATICS

The availability of genome sequence is just the beginning. Scientists now want to understand the genes and the role they play in the prevention, diagnosis and treatment of disease.

Randy Scott

OBJECTIVES

1. To understand the applications of bioinformatics in identifying genes
2. To upgrade knowledge about gene expression
3. To aid in understanding proteomics and drug design

INTRODUCTION

Without bioinformatics, new research in most fields of medicine and biology would come to a standstill. The explosion of publicly available genomic information resulting from the human genome project has precipitated the need for bioinformatics capabilities. This field of science is growing at an exponential pace with immense possibilities of innovation. Bioinformatics professionals are required to develop new tools with new algorithms that have better productive capacities and also apply the existing tools to gain new insights in molecular biology. Other areas of potential applications in vogue are

- Data storage and retrieval
- Annotation
- Analysis of genomic/proteomic/other high-throughput information
- Evolutionary model-building and phylogenetic analysis
- Architecture and content of genomes
- Complex systems analysis
- Genetic circuits
- Information content in DNA, RNA, protein sequence and structure
- Metabolic computing
- Data mining tools, neural nets, artificial intelligence
- Large-scale nucleic acid and protein sequence analysis

With this background, let us traverse through the next few pages of how knowledge of computers and biology has helped us take a giant leap in biomedical research.

Genomics is the study of an organism's entire genome. This field includes determining the entire DNA sequence of organisms and genetic mapping efforts. For the United States Environmental Protection Agency, the term "genomics" encompasses a broader scope of scientific inquiry and associated technologies than when it was initially considered. Thus, genomics is the study of all the genes of a cell, or tissue, at the DNA (genotype), mRNA (transcriptome), or protein (proteome) levels.[1]

Genomics was established by Fred Sanger when he first sequenced the complete genomes of a virus and a mitochondrion. His group established techniques of sequencing, genome mapping, data storage, and bioinformatic analyses in the 1970–1980s. A major branch of genomics is still concerned with sequencing the genomes of various organisms, but the knowledge of full genomes has created the possibility for the field of functional genomics, mainly concerned with patterns of gene expression during various conditions. The actual term "genomics" is thought to have been coined by Dr. Tom Roderick, a geneticist at the Jackson Laboratory (Bar Harbor, ME) at a meeting held in Maryland on the mapping of the human genome in 1986.

GENE IDENTIFICATION

Gene identification and functional classification start with the determination of coding sequence. Basically, an Open

Reading Frame (ORF) is determined from a start codon to a stop codon. This is relatively easy for prokaryotic genomes, where the gene density is high and introns are absent. Usually, ORFs longer than a certain threshold (300–500) are considered as potential genes. Genes that are longer than the threshold and genes on the opposite strand of the longer ORF (shadow genes) often lead to ambiguities which can be resolved by analysing the computational differences between coding regions, shadow genes and non-coding DNA. Predicting the coding sequences in eukaryotes is more difficult. The available gene-finding programs generate prediction on the basis of transcriptional signals (transcription start sites, TATA boxes, polyadenylation sites, etc.).

Translational signals (transcription initiation and termination sites) and splicing signals (donor and acceptor splice site positions). Among these programs are GENEMARK, GRAIL and GENEPARSER. These programs are continuously improved and more advanced programs like GENSCAN take into account reading frame compatibility of adjacent exons and compositional properties of introns and exons. Further increase in sensitivity can be obtained by including different sequence similarity functions for comparison to gene and protein sequences in available databases.

Eukaryotic Gene Prediction

Web servers for eukaryotic gene prediction on an "unknown" sequence will use coding prediction programs together with promoter, splice sites and poly(A) tail prediction programs. All the information gathered has to be combined to find the coding regions of the gene. There is no time to go into the theory behind the prediction methods.

Automatic gene prediction in higher eukaryotes does not work very well. Perhaps this is not surprising, considering differential and tissue-specific splicing, genes within genes, TATA-less promoters, the rare class of U12-dependent (AT-AC) splice sites and the need for species-specific parameters, all complicate the problem.

The sequences used for gene prediction may be obtained from any nucleotide repository, e.g. NCBI. Some examples of the gene prediction programs are Genemark, Genscan, Netgene2, HMM gene, etc. Other tools for specific needs are LBL promoter (to detect the promoters), TSSC prediction program (to find transcription factors), TSSW (to find translation factors) and POLYAH (to find the polyadenylation sites).

Prokaryotic Gene Prediction

The process of gene expression begins with transcription-the making of an mRNA copy of a gene by an RNA polymerase. Prokaryotic RNA polymerases are actually assemblies of several different proteins (alpha, beta and beta-prime) each of which plays a distinct and important role in the functioning of the enzyme. The –35 and –10 sequences recognized by any particular sigma factor are usually described as a consensus sequence—essentially the set of most commonly found nucleotides at the equivalent positions of other genes that are transcribed by RNA polymerases containing the same sigma factor.

Since stop codons are found in uninformative nucleotide sequences, approximately once every 21 codons (3 out of 64), a run of 30 or more triplet codons that does not include a stop codon is in itself a good evidence that the region corresponds to the coding sequence of a prokaryotic gene. One hallmark of prokaryotic genes that is related to their translation is the presence of the set of sequences around which ribosomes assemble at the 5′ end of each open reading frame. Often found immediately downstream of transcriptional sites and just upstream of the first start codon, ribosome loading sites (sometimes called Shine-Dalgarno sequences) almost invariably include the nucleotide sequence 5′AGGAGGU 3′.

Just as the RNA polymerases begin transcription at recognizable transcriptional start sites immediately downstream from promoters, the vast majority of prokaryotic operons also contain specific signals for the termination of transcription called intrinsic terminators. Intrinsic terminators have two prominent structural features; a sequence of nucleotides that include an inverted repeat and roughly six uracils immediately following the inverted repeats.[2]

GENOME ANNOTATION

Annotation is extraction of useful information from raw sequences, which includes identification of coding regions and genes in a genome and determination of what they do, a combination of comments, notations, references and citations, either in a free format or utilizing a controlled vocabulary which together describe the experimental and inferred information about a gene or a protein.

Scientists carry out annotations of sequences by using information linked to experiments. In contrast to sequences obtained through individual gene cloning experiments, in turn we should also be able to update annotations and apply to genes, the results of experimental analysis performed later on. This notion of integration is essential from a genomics point of view. Benchwork cannot follow the release pace of thousands of new potential genes and experimentally documented genes represent a minor fraction of genes. In order to compensate for this transitional lack of secure data, prediction tools are extensively used to analyse the sequences and extract putative information. By merging statistics, computer science and biological sciences, bioinformatics has developed many prediction tools. The analysis of the anonymous sequences combining different prediction programs and the results obtained by bioanalysts are the starting points of the annotation process. For this reason, the annotation work is mostly a predictive work and the result has to be considered as such.[3]

Whole genome annotation should not be taken as definitive and proven but rather as indication to help biologists in the sequence jungle and to derive future experimental approaches. This point is often forgotten when annotations are used. When in front of a large genomic sequence, the first problem is to localize all the genes on both the strands and more precisely, the different structural elements of these genes. This step is called the structural annotation, to clearly distinguish it from the following one, the functional annotation, which tries to find signs of function from the deduced protein sequences. Deep and detailed annotation implies numerous complementary analyses and checking. Unfortunately, because of the cost in time and money of this human expertise, genome annotation is generally restricted to the prediction of coding exons to deduce the protein sequence of potential genes and to label it with the functions of the closest homologue. We will compare and discuss the fast high-throughout annotation used in the systemic sequencing program and the possibilities of a deeper but slower annotation with two objectives—to highlight the dangerous traps when automatic annotations are blindly used and to present a few novel approaches and applications in genome annotation.

Structural Annotation

The very first information needed when analysing an unknown genomic DNA sequence is where the protein coding regions are located in this sequence. Gene structure prediction is not an easy task in prokaryotes, and is an even more difficult task in higher organisms, where genes are split into exons and introns, which can be difficult to detect. Modern gene detection software make use of artificial intelligence methods and are still far from being completely accurate, even for prokaryotic genes. A DNA sequence, which does not code for protein may still be important in the regulation of genes. They may contain recognition sequences for transcription factors. Identifying these patterns may provide a clue to the function of the sequence.

The prediction of the gene elements is a complex problem and its issue is primordial because of its consequences on all the following analysis. Eukaryotic genes with their mosaic structure are more difficult to find than prokaryotic ones, which are simple open reading frames. The presence of introns complicates the problem, although the binding sites of the spliceosomes may be used to predict the exact position of the exon borders. According to the prediction tools, the results of the prediction concern the splice sites, the exons or the whole gene (gene modelling software).

These prediction programs are based on two different approaches. The first version is called intrinsic based on the features of the genes and the genomes. Therefore significant and representative sets of only experimentally characterized genes are necessary to develop such efficient prediction programs. Furthermore, the origin of the training set has to be species-specific because each genome has its own feature and style. The second approach to find genes and exons, named extrinsic, uses the similarities detected in homologous genes (in general at the protein level to increase the sensibility of the search), or even better, identities with cognate transcript sequences. Only in this ideal last situation is the resulting gene structure asserted and not merely putative.

The performance (sensitivity and specificity) of the annotation software differ a lot, and consequently, the final predictions are not identically reliable. The structural annotation is generally semi-automatic because human intervention is necessary to integrate the results of the different predictions, in the final gene structure. Although a fully integrated annotation platform is still lacking, some annotation centres use an interactive method to visualize the output of the prediction tools and sequence similarity searches to help in this critical decision step.

Functional Annotation

At the present time, the functional genome annotation is based on the idea that some sequence similarities detected between two proteins means that they are homologous, i.e., they come from the same ancestor and share the same biochemical function. Therefore, for each predicted gene, the protein is deduced from the coding region and is compared through BLASTP with the protein databases. This minimum approach allows, in the best cases, the biochemical function of the gene product to be suggested. The high-throughout annotation realized by the annotation centres is too basic and quick to extract reliable information on the biological function. Some annotators confirm and complete the BLAST results by full-length alignment between the query protein and the closest homologue detected and by looking for motifs and family signatures.

This approach appears to be the best to attribute one or several biochemical functions to a predicted protein.

The name attributed to the predicted genes/proteins depends on the results of the homologous sequence search. Four categories of genes have been defined, but the associated nomenclature is not very homogeneous. The tendency is when a predicted gene product is 100 per cent identical to an already characterized protein; it receives the same name, whereas sequences with stringent similarity to known proteins are called putative proteins of the same name. The sequences for which only similarities to ESTs are detected are named unknown proteins. Finally, genes without similar sequences and hence only deduced from intrinsic prediction programs are labelled hypothetical.

GENE EXPRESSION

Gene expression is the process by which inheritable information from a gene, such as the DNA sequence, is made into a functional gene product, such as protein or RNA.

Several steps in the gene expression process may be modulated, including the transcription step and translation step and the post-translational modification of a protein. Gene regulation gives the cell control over structure and function, and is the basis for cellular differentiation, morphogenesis and the versatility and adaptability of any organism. Gene regulation may also serve as a substrate for evolutionary change, since control of the timing, location, and amount of gene expression can have a profound effect on the functions (actions) of the gene in the organism. Non-protein coding genes (e.g. rRNA genes, tRNA genes) are transcribed, but not translated into protein.[4]

DNA MICROARRAYS

DNA microarrays are small, solid supports onto which the sequences from thousands of different genes are immobilized or attached at fixed locations. The supports themselves are usually glass microscope slides—the size of two side-by-side fingers but can also be silicon chips or nylon membranes. The DNA is printed, spotted or actually synthesized directly onto the support.

DNA microarray revolutionizes the traditional way of one gene per experiment for the study of genomics in molecular biology. It allows thousands of genetic elements to be studied simultaneously by fabricating these genetic materials in an orderly fashion onto a single solid substrate. This massively parallel mechanism of genetic analysis enables thousands of genes to be monitored at a time and greatly increases the throughput for obtaining gene expression data.

Arrays of DNA can be spatially arranged, as in the commonly known gene chip (also called genome chip, DNA chip or gene array), or can be specific DNA sequences labelled such that they can be independently identified in solution. The traditional solid-phase array is a collection of microscopic DNA spots attached to a solid surface, such as glass, plastic or silicon biochip. Thousands of them can be placed in known locations on a single DNA microarray.

Microarray experiments rely on the principle of hybridization (base pairing). Instead of using only one probe, thousands of known probes are fabricated onto a solid substrate (e.g. glass slides) in a specified order, resulting in a DNA chip, or so-called microarray. After hybridization with the experimental sample (target), the microarray can be used to determine expression of genetic materials by examining the fluorescence present in each probe location.

An array is an orderly arrangement of samples. It provides a medium for matching known and unknown DNA samples based on base-pairing rules and automating the process of identifying the unknowns. An array experiment can make use of common assay systems such as microplates or standard blotting membranes, and can be created by hand, or can make use of robotics to deposit the sample. In general, arrays are described as macroarrays or microarrays, the difference being the size of the sample spots. Macroarrays contain sample spot sizes of about 300 microns or larger and can be easily imaged by existing gel and blot scanners. The sample spot sizes in microarray are typically less than 200 microns in diameter and these arrays usually contain thousands of spots. Microarrays require specialized robotics and imaging equipment.

Microarrays are significant both because they may contain a very large number of genes and because they are small in size. Microarrays are therefore useful when one wants to survey large number of genes quickly or when the sample to be studied is small. Microarrays may be used to assay gene expression within a single sample or to compare gene expression in two different cell types or tissue samples, such as in healthy and diseased tissues. A microarray can be used to examine the expression of hundreds or thousands of genes at once and so it promises to revolutionize the way scientists examine gene expression. This technology is still considered to be in its infancy; therefore, many initial studies using microarrays have represented simple surveys of gene expression profiles in a variety of cell types. Nevertheless, these studies represent an important and necessary step first in our understanding and cataloguing of the human genome.

As more information accumulates, scientists will be able to use microarrays to ask increasingly complex questions and perform more intricate experiments. With new advances, researchers will be able to infer probable functions of new genes based on similarities in expression patterns with those of known genes. Ultimately, these studies promise to expand the size of existing gene families, reveal new patterns of coordinated gene expression across gene families and uncover entirely new categories of genes. Furthermore, because the product of any one gene usually interacts with those of many others, our understanding of how these genes coordinate will become clearer through such analyses, and precise knowledge of these interrelationships will emerge. The use of microarrays may also hasten the identification of genes involved in the development of various diseases by enabling scientists to examine a much larger number of genes. This technology will also aid the examination of the integration of gene expression and function at the cellular level, revealing how multiple gene products work together to produce physical and chemical responses to both static and changing cellular needs.

TOOLS FOR GENE EXPRESSION ANALYSIS

- Avadis
- BiNGO
- CLENCH
- Database for Annotation, Visualization and Integrated Discovery (DAVID)
- EASE
- EasyGO
- eGOn V2.0 (explore Gene Ontology)
- ermineJ
- FatiGO
- Functional Information Viewer and Analyzer (FIVA)
- FuncAssociate
- FuncExpression
- FunCluster, Functional Profiling of Microarray Expression Data
- FunNet: Functional Analysis of Transcriptional Networks
- G-SESAME
- GARBAN
- GENECODIS
- GeneMerge

- GFINDer: Genome Function INtegrated Discoverer
- GOALIE (Generalized Ontological Algorithmic Logical Invariants Extractor)
- GOArray
- GOdist
- Gene Ontology Enrichment Analysis Software Toolkit (GOEAST)
- GOHyperGAll
- High-Throughput GoMiner
- L2L
- Meta Gene Profiler (MetaGP)
- MultiExperiment Viewer (MeV)
- Probe Explorer
- ProfCom, Profiling of Complex Functionality
- SeqExpress
- SerbGO
- Stanford Microarray Database
- Spotfire Gene Ontology Advantage Application
- Short Time-series Expression Miner (STEM)
- T-Profiler
- THEA

APPLICATIONS

The use of microarrays have made research easier and more relative of late. Given below are the key areas of research.

Diagnostics

DNA array technology provides a method for rapid genotyping, facilitating the diagnosis of diseases for which a gene mutation has been identified. It also assists in the diagnosis of diseases for which known gene expression biomarkers of a pathologic state or signature genes, exist. Signature genes are genes that are constitutively expressed in a normal or diseased cell or tissue that can serve as an identifier of that cell or tissue.[5]

Custom Drug Selection

The presence of alternate gene forms or atypical expression of a gene involved in drug action or metabolism can manifest as resistance to therapy, or an atypical response to therapy. Studies correlate with the genetic profile of individual patients and the individual response to a drug or toxin, respectively. The information obtained from these studies can be used to design arrays that will assist in the selection of custom and rational drug therapy.

Discovery of Therapeutic Targets

The identification of signature genes or biomarkers indicative of a disease process can identify candidate targets for therapeutic intervention.

Predict Drug/Toxin Activity

Arrays assist in the identification of sentinel genes that demonstrate altered expression in a given cell or tissue type in response to a drug or toxin exposure. Creating profiles of sentinel genes associated with drugs, sharing a common mechanism allows potential new therapies to be rapidly screened for similar activities. This facilitates the selection of compounds for further investigation and may reduce the need for animal testing. An *in vitro* screen for potential toxicity has the potential to reduce drug-screening costs, prevent human suffering and reduce product liability.

DNA array analysis allows multiple sequences to be searched for the presence of a suitable biomarker or a group of biomarkers. Testing the effects of a drug on a group of biomarkers may compensate for differences that reflect genetic variability.

Determination of Pharmacological Mechanism

Analysis of essential genes can assist in determining the mechanism of action of a drug or toxin. Given that there is a multitude of events triggered by the initial action of a drug, screening thousands of genes at one time can identify multiple potential drug effectors. This allows robust hypotheses of drug mechanism to be formed and tested.

SOFTWARE TOOLS

Many tools are available for various purposes in bioinformatics. Described below are a few important tools for basic use.

Glimmer

A gene finder derived from Glimmer but developed specifically for eukaryotes. It is based on a dynamic programming algorithm that considers all combinations of possible exons for inclusion in a gene model and chooses the best, of these combinations. The decision about what gene model is best is a combination of the strength of the splice sites and the score of the exons generated by an Interpolated Markov Model (IMM). The system has been trained for *Arabidopsis thaliana, Oryza sativa (rice)* and *Plasmodium falciparum* (the malarial parasite) and should work well on closely related organisms. We have a system for finding genes in the microbial DNA—especially the genomes of bacteria and archaea (Gene Locator and Interpolated Markov Modeller) which use Interpolated Markov models to identify the coding regions and distinguish them from non-coding DNA.

Gene Splicer

This is a fast, flexible system for detecting splice sites in the genomic DNA of various eukaryotes. The system has been trained and tested successfully on *Plasmodium falciparum, Arabidopsis thaliana* and human genomes. Training data sets for human and *Arabidopsis thaliana* are included.

Trans Term

TransTerm is a program that finds Rho-independent transcription terminators in bacterial genomes.

Each terminator found by the program is assigned a confidence value that provides an estimate of its probability of being a true terminator.

RepeatFinder

RepeatFinder is a computational system for analysis of repetitive structures of genomic sequences. The method uses suffix trees for efficient computation of exact repeats and organizes those repeats into classes. The method can be applied to individual genome sequences or sets of sequences. The output is multi-FASTA file of the found repeat sequences that can be used as the target of searches.

RBS Finder

RBS finder is a Perl script that implements an algorithm to find ribosome-binding sites for genes in bacterial and archaeal genomes. It is normally run as a post-processor to the Glimmer gene finder or to other prokaryotic gene finders.

TIGR Combiner

The TIGR Combiner uses a voting scheme to combine the predictions of 3 or more gene finders and produce a single best prediction. It is compatible with GlimmerM, Genescan, FGenes, GRAIL and GeneMark.HMM.

MICROARRAY DATABASES

Microarray databases fulfil different functions and can be classified as follows:

1. Those that provide public access to database software for local installation
2. Those for public data deposition
3. Those available for public queries

Given below are some useful microarray resources.

ArrayExpress http://www.ebi.ac.uk/microarray-as/ae/

Database of gene expression and other microarray data at the European Bioinformatics Institute (EBI); public data deposition and public queries.

BASE http://base.thep.lu.se/

BASE (BioArray Software Environment) is a comprehensive free web-based database solution for the massive amounts of data generated by microarray analysis-local installation.

ChipDB http://staffa.wi.mit.edu/chipdb/public/index.html

A searchable database of gene expression; Centre for Genome Research public queries.

Dragon http://pevsnerlab.kennedykrieger.org/dragon.htm

Database referencing of array genes online; public queries.

ExpressDB http://arep.med.harvard.edu/ExpressDB/

A relational database containing yeast and *E. coli* RNA expression data; public queries of *E. coli* and yeast data.

Expression Connection http://db.yeastgenome.org/cgi-bin/expression/expressionConnection.pl

To search gene expression data from multiple microarray studies; public queries of yeast data.

GeneDirector http://www.biochipnet.com/node/2954

A comprehensive microarray data management solution; local installation.

GeneX public http://genex.sourceforge.net/

An open source database for storing, processing, and analyzing gene expression microarray data; local installation and public data deposition.

GEO http://www.ncbi.nlm.nih.gov/geo/

NCBI's gene expression/molecular abundance repository; public data deposition and public queries.

GXD http://www.informatics.jax.org/mgihome/GXD/aboutGXD.shtml

Gene expression database; data deposition and public queries of mouse data.

mAdb http://nciarray.nci.nih.gov/

NCI's microarray data management and analysis system;local installation.

NOMAD http://ucsf-nomad.sourceforge.net/

A public source for microarray protocols and software maintained by UCSF; local installation.

RAD http://www.cbil.upenn.edu/RAD/php/index.php

A resource for gene expression studies; public queries.

SMD http://genome-www5.stanford.edu/

Stanford Microarray Database; local installation and public queries.

yMGV http://transcriptome.ens.fr/ymgv/

Yeast Microarray Global Viewer (yMGV) is an on-line database providing a synthetic view of the transcriptional expression profiles of yeast genes; public queries of yeast data.

BIOINFORMATICS CAREERS

Career opportunities for bioinformatics professionals are available in various streams. Given below is a non-exhaustive list of where one might find oneself after obtaining a focused training in any of the many fields of bioinformatics.

- Gene Analyst
- Protein Analyst
- Phylogenetist
- Research Scientist/Associate

- Database Programmer
- Bioinformatics Software Developer
- Computational Biologist
- Network Analyst
- Structural Analyst
- Molecular Modeller
- Bio-Statistician
- Biomechanic
- Cheminformatician
- Pharmacogenetician
- Pharmacogenomic Research Scientist

REVIEW QUESTIONS

1. Highlight the differences between prokaryotic and eukaryotic genes.
2. What are microarrays? Discuss the methodology and applications.
3. Enlist the applications of the proteomics.

REFERENCES

1. EPA Interim Genomics Policy.
2. James, W. Fickett. (1998). "Gene identification." *Bioinformatics*. 10: 563–578.
3. Shoemaker, D.D., Schadt, E.E., Armour, C.D., He, Y.D., Garrett-Engele, P., McDonagh, P.D. and Loer, P.M. (2001). "Experimental annotation of the human genome

using microarray technology." *Nature*. Vol. 409 No. 6822, 922–927.

4. Lockhart, D.J. and Winzeler, E.A. (2000). "Genomics, gene expression and DNA arrays." *Nature*. 405(6788):827–836.

5. Wallace, R.W. (1997). "DNA on a chip-serving up the genome for diagnostics and research." *Molecular Medicine Today*. 3: 384–389.

APPENDIX

BIOINFORMATICS-RELATED WEBSITES

htto://www.wileyxorn/legacy/prcKJucts/subjecMife/bioinformatic5/courses.html

http://www. iscb.org/univ.shtml

http://bioinf.man.ac.uk/ember/PDF/CALreport.pdf

http://www.crust.bbk.ac.uk/PPS2/course/index.html

http://www.cmu.edu/biQ/education/masters.html

http://www.cbcb.duke.edu//

http://icg.harvard.edu/~bphysl0l/

http://www.indiana.edu/%7Echeminio/401svlOQ.html

http://www.ihu.edu/~ibr/bwf.html

http://Qenomics.ncsu.edu/bioinfo.html

http://genomics.ncsu.edu/function.html

http://scpd.stanford.edu/SCPD/courses/proEd

http://www.smi.stanford.edu/proiects/heli^mi214/

http://cmgm.stanford.edu/biochem218/#Description

http://titan.biotec.uiuc.edu/

http://bioinfo.med.unc.edu/Bioinformatics/index.html

http://genome.ucsd.edu/

http://www.sdsc.edu/pb/Edu.html

http://bioinformatics.ucsd.edu/

http://www.cse.ucsc.edu/centers/cbe/prop-ms.html

http://www.soe.ucsc.edu/research/compbio/

http://www.dcs.ex.ac.uk/~anarauan/bioinf/msc.htm

http://www.pcbi.upenn.edu/educatton.php3

http://monod.uwaterloo.ca//courses.php3

http://www.vir oloay.wisc.edu/acp/pases/Classes/ClassMaterials/ TableofWords_711_Q0.html

http://www.cibm.wisc.edu/

http://www.Aladdinsus.com/expander/

http://www.adobe.com/prodindex/acrobat/readstep.html

http://www.macromedia.com/shockwave/download/

http://www.umass.edu/microbio/rasmol/qetras.htm

http://www.mdli xom/support/chime/ownload.html

http: //www, umass. edu/microbio/ch ime/index. html

http://www.apple.com/quicktime/install/

http://www.real.com/products/player

http.//www.microsoft.com/windows/mediaplayer/

http://www.microsoft.com/macoffice/productinfo/98dl/pptvdl.htm

http://officeupdate.microsoft,com/downloadDetails/ppviewl6.htm

http://officeupdate.microsoft.com/downloadDetails/ppyiew97.htm

http://www.accessexcellence.org/index.html>

http://www.accessexcellence.org/AB/BC/

http ://www.accessexcellence.org/AB/WYW/index.html

http://ndbserver.rutgers.edu/NDB/archives/NAintro/index.html

http://www.res. bbsrc.ac.uk/molbio/guicie/

http://locutus.lsic.ucla.edu/ls3/tutorials/

http://www.cbc.umn.edu/~mwd/cell.html

http://gened.emc.maricopa.edu/bio/biol81/BlOBK/BioBookTOC.html

http://esg-www.mit.edu:8001/esgbio/

http://www.ornl.gov/TechResources/Human_Genome/publicat/primer/intro.html

http://cti.itc. Virginia. edu/~cmg/

GLOSSARY

Allelomics Study of a collection of different allotypes or allelic protein variants, a new type of protein library.

Allergenomics Rapid and comprehensive analysis of putative proteinous allergens (allergenome), by applying such a proteomic strategy.

Bibliomics A subset of high quality and rare information, retrieved and organized by systematic literature-searching tools from existing databases, and related to a subset of genes functioning together in "-omic" sciences.

Biomics The oldest "-ome" suffix series. It was coined in 1916. It is a study that refers to an ecological community of organisms and environments. The ability of genes or alleles to affect the representation of the host organism in a biome is an operational definition for the "function" of the gene.

Cellomics Study of cell function and drug impact at the level of the cell.

Chemoproteomics The use of biological information to guide chemistry offers a highly efficient alternative to small-molecule characterization that can accelerate drug discovery.

Chromosomics The term "chromosomics" is introduced to draw attention to the three-dimensional morphological changes in chromosomes, that are essential elements in gene regulation. It deals with the plasticity of chromosomes in relation to the three-dimensional positions of genes, which affect cell function in a developmental and tissue-specific manner during the cell cycle.

Chronomics Technology that allows the monitoring of ever denser and longer serial biological and physical environmental data. This in turn allows the recognition of time structures, chronomes, including, with an ever broader spectrum of rhythms, also deterministic and other chaos and trends.

Computational RNomics Versatile and reliable computational methods that can detect and classify functional RNAs, preferably within a single genome, or in case this proves impossible, from a very small set of related genomes.

Cryobionomics A novel term describing the remodelled concept of genetic stability and the reintroduction of cryo-preserved plants into the environment.

Crystallomics Study of the production of highly purified protein samples and diffraction quality crystals.

Cytomics Multiparameter cytometric analysis of the cellular heterogeneity of cytomes, to access maximum information on the apparent molecular cell phenotypes, resulting from cell genotypes and exposure. Molecular cell phenotypes is the naturally existing cellular and cell population heterogeneity of disease affected body. Cytomes contain the information on the future development (prediction) as well as on the present status (diagnosis) of a disease.

Degradomics The application of genomic and proteomic approaches to identify the protease and protease-substrate repertoires, or "degradomes", on an organism-wide scale—promises to uncover new roles for proteases *in vivo*. This knowledge will facilitate the identification of new pharmaceutical targets to treat disease.

Diagnomics Molecular diagnostics with highly enhanced prognostics, diagnostics and therapeutic benefits combined with advanced genomic and proteomic technologies.

Embryogenomics Developmental biology meets genomics.

Enviromics The study of the total complement of environmental characteristics, conditions, and processes required for life form viability and successful adaptation. The genome dwells within an environment and genomic expression shapes and is shaped by environment.

Epigenomics A whole genome approach to epigenesis and epigenetics.

Epitomics A new field of science that studies all epitopes of the proteome in an organism. The understanding of epitope reveals the functions of these proteins.

Ethnogenomics To study the characteristics of genomic polymorphism and genomic diversity of various groups of population: separate communities, ethnoses, and ethnoterritorial communities.

Expressomics Study of transcripts, proteins and other ligands. It has two major branches. One is RNA expression represented by transcriptomics and the other is protein expression by proteomics (or translatomics).

Fluxomics Integration of metabolic pathway engineering and fermentation production.

Fragmentomics Study of fragments of proteins that are the true biomarkers.

Fragonomics Study of smaller molecules (fragments) in the drug discovery process has led to success in delivering novel leads for many different targets.

Functional lectinomics Encompasses, among other activities, intra- and intercellular transport processes, sensor branches of innate immunity, regulation of cell-cell (matrix) adhesion or migration and positive/ negative growth control with implications for differentiation and malignancy.

Functionomics Used as a synonym for functional genomics.

Functomics A comparison of annotation schemes for genomes.

GPCRomics Defined as the application of a wide range of technologies to study the GPCR (GProtein-coupled receptor) family of proteins which has traditionally provided the pharmaceutical industry with a rich source of targets for drug discovery.

Hygienomics The study developed to assist management in their introduction of hygiene and food safety systems. For an effective introduction, the management systems must be designed to fit with the current generational state of an organization.

Immunomics Study of the molecular functions associated with all immune-related coding and non-coding mRNA transcripts.

Inomics The study of inositol signalling molecules (ISMs) and their role in regulating cellular functions.

Interactomics Study of a complete set of macromolecular interactions along with physical and genetic factors.

Invariomics The study of the complement of genes in an organism whose level of expression does not change significantly from one condition to another, i.e., they are invariantly expressed.

Ionomics The use of mineral nutrient and trace element profiling as a new tool to determine the biological significance of connections between a plant's genome and its elemental profile.

Kinomics Study of the protein kinases.

Ligandomics The study of the complete set of organic small molecules.

Lipidomics A rapidly expanding research field in which multiple techniques are utilized, to quantitate the hundreds of chemically distinct lipids in cells and determine the molecular mechanisms through which they facilitate cellular function.

Mechanomics The study of the mechanical systems at work within an organism.

Metabolomics The study of the metabolite profiles in biological samples, is growing amidst the current shift toward translational research.

Metabonomics The quantitative measurement of the dynamic multiparametric metabolic response of living systems to pathophysiological stimuli or genetic modification. This concept has arisen from work on the application of ^{1}H-NMR spectroscopy to study the multicomponent measurement of biofluids, cells, and tissues.

Metallomics The study of the entirety of the content of inorganic species within a cell or tissue type.

Metaproteomics The large-scale characterization of the entire protein complement of environmental microbiota at a given point in time.

NObonomics The global profiling of NO metabolism and signalling *in vivo.*

Operomics The profiling of tissues and cell populations at the genomic, transcriptomic and proteomic levels.

Pharmacophylogenomics Study of genes, evolution and drug targets. *See also* phylogonomics.

Phylogenomics A new discipline that concentrates on large-scale analysis on the whole genome level of phylogeny prediction. Pharmacophylogenomics attempts to apply this knowledge to pharmacological relevant areas like for example, *ADME*-Tox (**A**bsorption, **D**istribution, **M**etabolism, **E**xcretion and **T**oxicology).

Phenomics Study of phenotypes with knowledge of the genotypes.

Physiomics Knowledge of the complete physiology of an organism, including all interacting metabolic pathways, structural and biochemical scaffolding, the proteins and accessories that make them up, and the gene interactions and cues that control them.

Physionomics The term proposed for the comprehensive physiological profiling of the plant system, following the parallel terminology of the molecular and biochemical "-omics" technologies. Physionomics procedures provide a first clue to the mode of action of a new herbicide that can direct more time consuming and costly molecular, biochemical, histochemical or analytical studies to identify a target site more efficiently.

Post-translatomics Various protein modifications finely tune the cellular functions of each protein. Understanding the relationship between post-translational modifications and functional changes is "post-translatomics".

Promoteromics Study of the complete set of promoters.

Proteogenomics The study of gene expression during the infectious cycle, in mutants or after environmental or chemical stimuli, is a powerful approach towards understanding parasite virulence and the development of control measures.

Pseudogenomics Study of the complement of pseudogenes in the proteome.

Regulomics The study of the totality of specific molecular interactions that determine gene expression in any given organism, and includes the topological (circuitry) characteristics of the interaction networks as well as the quantitative variations of their components.

Resistomics Study of all those proteins whose expression is altered in drug resistant forms.

RNomics The understanding of functional RNAs and their interactions at a genomic level.

Secretomics A study of a subset of the proteome that is defined by its action, i.e., it is actively exported from the cell.

Separomics To provide a simple solution to every purification situation.

Somatonomics Study of all somatic gene rearrangements, lymphoid plus non-lymphoid.

Targetomics Transitioning from the identification to the subsequent validation and prioritization of their cognate proteins as bonafide drug targets using proteomic techniques—a process that could appropriately be termed targetomics.

Transcriptomics The study of the population of mRNA transcripts in the cell, weighted by their expression levels.

Transductomics Study of signalling pathways.

Transportomics The study of the population of the gene products that are transported; this includes the secretome.

Vaccinomics Using bioinformatics and genomics for vaccine development.

Variomics Study of variants of DNA, RNA and proteins.

Viromics The use of viruses and viral gene transfer to explore the complexity arising from the vast array of new targets available from the human and murine genomes.

INDEX

C

D

E

F

G

H

U

V

W

X

Y

Z

www.ingramcontent.com/pod-product-compliance
Ingram Content Group UK Ltd.
Pitfield, Milton Keynes, MK11 3LW, UK
UKHW021913290826
14793UKWH00003B/239

9 788180 943867